The Collected Courses of the Academy of European Law
Series Editors: Professor Marise Cremona
Professor Bruno de Witte, and
Professor Francesco Francioni,
*European University Institute,
Florence*
Assistant Editor: Anny Bremner, *European University
Institute, Florence*

VOLUME XVIII/3
Environmental Protection
European Law and Governance

The Collected Courses of the Academy of European Law
Edited by Professor Marise Cremona,
Professor Bruno de Witte, and Professor Francesco Francioni
Assistant Editor: Anny Bremner

This series brings together the Collected Courses of the Academy of European Law in Florence. The Academy's mission is to produce scholarly analyses which are at the cutting edge of the two fields in which it works: European Union law and human rights law. A 'general course' is given each year in each field, by a distinguished scholar and/or practitioner, who either examines the field as a whole through a particular thematic, conceptual, or philosophical lens, or who looks at a particular theme in the context of the overall body of law in the field. The Academy also publishes each year a volume of collected essays with a specific theme in each of the two fields.

KJE 27 ENV

Environmental Protection

European Law and Governance

Edited by
JOANNE SCOTT
Professor of European Law, University College London

UNIVERSITY PRESS

OXFORD
UNIVERSITY PRESS

Great Clarendon Street, Oxford OX2 6DP

Oxford University Press is a department of the University of Oxford.
It furthers the University's objective of excellence in research, scholarship,
and education by publishing worldwide in

Oxford New York

Auckland Cape Town Dar es Salaam Hong Kong Karachi
Kuala Lumpur Madrid Melbourne Mexico City Nairobi
New Delhi Shanghai Taipei Toronto

With offices in

Argentina Austria Brazil Chile Czech Republic France Greece
Guatemala Hungary Italy Japan Poland Portugal Singapore
South Korea Switzerland Thailand Turkey Ukraine Vietnam

Oxford is a registered trade mark of Oxford University Press
in the UK and in certain other countries

Published in the United States
by Oxford University Press Inc., New York

© The several contributors, 2009

The moral rights of the authors have been asserted
Crown copyright material is reproduced under Class Licence
Number C01P0000148 with the permission of OPSI
and the Queen's Printer for Scotland

Database right Oxford University Press (maker)

First published 2009

All rights reserved. No part of this publication may be reproduced,
stored in a retrieval system, or transmitted, in any form or by any means,
without the prior permission in writing of Oxford University Press,
or as expressly permitted by law, or under terms agreed with the appropriate
reprographics rights organization. Enquiries concerning reproduction
outside the scope of the above should be sent to the Rights Department,
Oxford University Press, at the address above

You must not circulate this book in any other binding or cover
and you must impose the same condition on any acquirer

British Library Cataloguing in Publication Data

Data available

Library of Congress Cataloging in Publication Data

Environmental protection : European law and governance / edited by Joanne Scott.
  p. cm. — (Collected courses of the Academy of European Law)
Includes bibliographical references and index.
ISBN 978–0–19–956517–7
 1. Environmental Law—European Union countries. 2. Environmental policy—European
Union countries. 3. Environmental protection—European Union countries. 4. Nature
conservation—Law and legislation—European Union countries. 5. Environmental impact
charges—Law and legislation—European Union countries. I. Scott, Joanne.
  KJE6242. E588 2009
  344.2404'6—dc22   2008048916

Typeset by Newgen Imaging Systems (P) Ltd., Chennai, India
Printed in Great Britain
on acid-free paper by
The MPG Books Group in the UK

978–0–19–956517–7

1 3 5 7 9 10 8 6 4 2

Preface

Joanne Scott

This volume examines the past and present of EU environmental law and governance, and contemplates its future. The chapters which follow range from the general to the particular, and span both the internal and external dimensions of EU law and policy on the environment. In certain respects the chapters read as detective stories without conclusions. The instruments which they describe provide frameworks for achieving social and environmental change. But the implementation phase remains all important. There is room for reasonable disagreement about the appropriateness and adequacy of the frameworks put in place. Ultimately though, the proof will be in the pudding, and none of us can yet claim definitive knowledge about the shape that this will take. What is clear is that the EU has sought to put in place frameworks which encourage the generation of rich, up-to-date, information about environmental quality and about the hazards which threaten this. Thus, the frameworks are themselves designed to generate the informational resources necessary for policy evaluation, and for policy adjustment in the light of the lessons learned. It is also the case that the direct involvement of environmental NGOs in some of the implementation processes described can serve to facilitate their informed and critical engagement with the nature of those processes and with the outcomes which ensue from them. An NGO 'technical review' on the implementation of the Water Framework Directive served both to attest to insufficiencies in implementation when it comes to the key task of defining 'good water status' but, also more positively, to illustrate the key role played by such environmental groups in promoting on-going accountability in implementation.[1]

The volume begins with a chapter by Ingmar von Homeyer. This constitutes the 'real' introduction to this volume. It provides a sweeping overview of EU environmental governance. It argues that it is possible to distinguish four governance phases, which von Homeyer labels as environmental, internal market, integration, and sustainable development. Following an analysis of the factors affecting the evolution of EU environmental governance, von Homeyer leads us through the four governance phases, distinguishing these on the basis of a variety of characteristics. Among the characteristics discussed are decision-making mode and instruments used. In terms of decision-making mode, for example, von Homeyer suggests that we have moved from a technocratic approach associated

---

[1] See, for a brief overview, <http://www.eeb.org/activities/water/20061010-NGO-Policy-Brief-on-WFD-Intercalibratiob.pdf> and the discussion in Maria Lee's contribution to this volume.

with the original environment regime, to a sustainable development approach which emphasizes the importance of continuous learning in multi-levelled and often participatory policy spaces. In terms of instruments, to take a second example, it is suggested that the EU has moved from legally binding, top-down, regulation, to greater reliance on framework directives which place additional emphasis upon the implementation phase. Von Homeyer argues, in keeping with the learning dimension, that these instruments are characterized by reflexivity in that they support and encourage comparison of performance across Member States, and mutual learning on the basis of the benchmarks derived. For von Homeyer the Water Framework Directive, discussed in detail in Lee's chapter to follow, is emblematic of this latest governance phase. While von Homeyer uses a temporal perspective, he is careful to emphasize the phenomenon of institutional layering. According to this, governance regimes may co-exist over long periods of time, and interact in various ways. Von Homeyer concludes by considering some of the reasons for this. He also reminds us that his story is not yet complete, and speculates that we could be seeing the beginning of a new governance phase, associated in particular with climate change, and marking the comeback of a more command and control based approach.

Maria Lee's chapter is concerned with the Water Framework Directive as an instance of 'new governance' in the EU. It adopts an approach which is constructively critical. The chapter begins with an overview of this 'dauntingly complex' instrument. It then outlines and explores the principal governance techniques on which it relies. Much attention is paid to the fascinating Common Implementation Strategy which has emerged organically as a forum for multi-level, multi-actor, collaboration in implementation. While Lee is clear that 'new governance allows the pursuit of objectives that simply could not have found a home in legislation', she is appropriately demanding in her scrutiny of the Common Implementation Strategy. She asks hard and important questions about accountability in this context, highlighting the important role of experts in this process, and evaluating the practice of, and the potential for, meaningful public participation. As with so many of the chapters in this volume, she leaves us in a tantalizing space. There is much promise, but also room for concern. An evaluation of EU environmental law and governance is not a static process, but must evolve along the way as implementation proceeds. As the Water Framework Directive itself exemplifies, this implies an important and demanding role for non-governmental organizations in the environmental domain.

My own chapter examines the EU's new chemicals regime, known by the acronym REACH. This puts in place a new regulatory framework for both new and existing chemicals, and concerns the Registration, Evaluation, Authorization and Restriction of Chemical substances. This seeks to promote a high level of protection of human health and the environment, in part by promoting the replacement of risky substances with less risky alternatives. At the same time it seeks to contribute to the smooth functioning of the internal market, in a heavily traded product sector. The chapter highlights certain core elements of this complex instrument, including crucially the informational burden which it places on industry. This regulation creates a system of governance which is intensely fractured. Power is shared among a multiplicity of actors, operating at different levels of government, in the private and public spheres. No single actor has autonomous decision-making power. Rather, each is empowered, in different circumstances and in different ways, to play a role in maintaining the dynamic quality of regulation in the face of information deficits and uncertain risks. Different actors can play a role in seeking to ensure the continuous generation of new and better information about risk and about the mitigation of risk, and in seeking to prompt regulatory decisions which are appropriately responsive to this. This fracturing of power creates a governance framework which is complex. But it is also a framework which seeks to combine, in a novel way, harmonization with dynamism, and uniformity with structures for regulatory learning.

In her chapter on Europe's emerging spatial policy, Jane Holder's starting point is the profound impact that a range of EU policies have had on land use patterns and landscapes in the EU. She offers an analysis of the European Spatial Development Perspective, which provides a framework for decision-making regarding land use development in the EU. Holder describes and analyses the elements of this, and highlights the governance techniques which characterize both the adoption and implementation processes. It becomes apparent that the implementation phase is the more inclusive one, in terms of the breadth of participation which it invites. Following the now familiar logic of 'new governance' approaches, she points out the way in which the strategy uses information generation and exchange, together with the adoption of common criteria and indicators, as techniques for governance in this sphere. Holder stresses that though the Spatial Perspective is committed to ensuring evenly distributed economic growth throughout the territory of the EU, there is a conspicuous absence in relation to considerations of environmental justice. This she defines as being concerned with equal access to a clean environment and equal protection from possible harm irrespective of race, income, class, or any other distinguishing feature of socio-economic status. Holder surveys the scholarship growing up around the spatial dimension of EU policy, and argues that it marks the beginning of a 'critical geography of law, land and power in the EU'.

Massimiliano Montini's chapter examines the legal issues surrounding the European Union's competence in relation to environment. His focus is on external relations competence though, as he observes, the line separating internal from external can be hard to draw, given the profound effects of much 'internal' environmental law on states and other actors operating outside of the territory of the EU. Montini provides an overview of the general picture in external relations competence, express and implied, before turning to the specific theme of environmental policy. In keeping with the case study approach of this volume, Montini presents two case studies, one broad and the other more specific. The first case study concerns the trade and environment interface. Montini explores the WTO framework within which the EU operates, and the constraints this implies for the EU in promulgating trade related environmental measures. The concept of trade-related environmental measures is now extremely broad, in view of the breadth of the concept of a technical regulation in the Technical Barriers to Trade Agreement (TBT Agreement), and also the scope of application of the Sanitary and Phytosanitary Measures Agreement (SPS Agreement). Montini's second case study concerns EU participation in the international regulatory framework for climate change; a topical theme also addressed in the final chapter by Lefevere. Montini examines the status of the EC in this international regime, and addresses the issues of legal responsibility arising. One focus of this second case study is the EU's burden-sharing arrangement under Kyoto. Because of this, the EU has been able to divide the burden of compliance with Kyoto unevenly as between the different EU Member States, in accordance with the notion of common but differentiated responsibility. The extra flexibility which this permits the EU is indicative, in subsidiarity terms, of the value added of EU level action in the international domain.

The chapter by Jürgen Lefevere offers a fascinating account of recent developments in the global management of climate change. Though written from a personal perspective, it offers an insider's account of progress to date and of the challenges ahead. The governance dimension is never far from the surface. Be it in relation to the role of politics and experts in the Intergovernmental Panel on Climate Change, or in relation to the Gleneagles Action Plan and Dialogue, the chapter exemplifies the multiplicity of players on the global climate change stage, and the variety of institutional forums for combining or coordinating their efforts. Here, as elsewhere, the chapter highlights and evaluates the stance and role of the EU in the global governance of climate change. It examines EU participation in the work of the IPPC, as well as the 'mainstreaming' of climate change in an integrated EU policy on climate change, energy, and competitiveness. Lefevere also highlights the mainstreaming challenge at the international level including, for example, in trade and finance, and in the UN system. The chapter concludes with a close look at the Bali UN climate change conference in late 2007, and with a critical look at the Bali Action Plan, including from the perspective of the EU. Lefevere identifies the key issues facing climate change negotiators in the coming years. Among the most challenging is that concerning the mobilization of finance and investment for climate change mitigation and adaptation. On a pessimistic note, and drawing on his own experience, Lefevere confesses that during his 'ten years as a climate negotiator', he has 'increasingly felt that we were discussing the wrong issues among the wrong people', and hence that the policy mainstreaming challenge is far from having being met.

These chapters have their origins in the Academy of European Law taught at the European University Institute in Florence in July 2007. They took shape in the course of stimulating discussions with a large and lively group of students and scholars from a wide variety of countries and walks of life. I would like, on behalf of all of the contributors, to thank the Directors of the Academy for their invitation to participate, and the participating students and scholars for the excellence of their input. Particular thanks are due also to Anny Bremner for her first rate assistance before, during, and after the Academy. Lucy Page has been a constructive and patient editor at OUP, and we owe her our thanks.

Summary Contents

List of Contributors	xvii
Table of Cases	xix
Table of Legislation	xxi
1. The Evolution of EU Environmental Governance Ingmar von Homeyer	1
2. Law and Governance of Water Protection Policy Maria Lee	27
3. REACH: Combining Harmonization and Dynamism in the Regulation of Chemicals Joanne Scott	56
4. Building Spatial Europe: An Environmental Justice Perspective Dr Jane Holder	92
5. EC External Relations on Environmental Law Massimiliano Montini	127
6. A Climate of Change: An Analysis of Progress in EU and International Climate Change Policy Jürgen Lefevere	171
Index	213

Contents

List of Contributors	xvii
Table of Cases	xix
Table of Legislation	xxi

1. The Evolution of EU Environmental Governance — 1
  1. Introduction — 1
  2. Factors Affecting the Evolution of EU Environmental Governance — 4
    A. Environmental and Economic Conditions — 4
    B. International Political Developments and Commitments — 4
    C. Major Changes of the EU Polity — 5
  3. EU Environmental Governance Regimes — 7
    A. The Environment Regime — 8
    B. The Internal Market Regime — 11
    C. The Integration Regime — 14
    D. The Sustainable Development Regime — 18
  4. Conclusion — 24

2. Law and Governance of Water Protection Policy — 27
  1. Introduction — 27
  2. The Water Framework Directive — 29
  3. Flexibility, Decentralization, and Control — 36
  4. The Substantive Filling Out of the Water Framework Directive: Governance Beyond the Legislation — 45
  5. Conclusions — 55

3. REACH: Combining Harmonization and Dynamism in the Regulation of Chemicals — 56
  1. Introduction — 56
  2. Core Element: Industry Responsibility — 60
    A. Registration — 60
    B. Applications for Authorization — 66
  3. Core Element: Contestability — 68
    A. Substance Evaluation — 69
    B. Substances Requiring Authorization — 69
    C. Restrictions — 71
    D. Harmonized Classification and Labelling — 71
    E. Member State Safeguards — 72

F. Article 95 EC   72
   G. Contestability: The Limitations   73
   H. A Note on Authorization and Restriction Procedures
      and Criteria   75
4. Core Element: Substitution 78
5. Core Element: Provisionality 80
   A. Review of Authorizations   80
   B. Reporting/Review/Revision   82
6. Core Element: Transparency 84
   A. Information in the Supply Chain   84
   B. Classification and Labelling of Substances   86
   C. Access to Information   89
7. Conclusion 90

4. **Building Spatial Europe: An Environmental Justice Perspective** 92
   1. Introduction 92
   2. Bringing in Space in Europe 94
      A. The European Spatial Development Perspective   98
      B. Elements   105
   3. Absence: Environmental Justice 118
   4. Conclusions 124

5. **EC External Relations on Environmental Law** 127
   1. Introduction 127
   2. EC External Relations and the Protection of the Environment   128
      A. Introduction on the Existence and the Exclusivity Question   128
      B. The Existence Question   129
      C. The Exclusivity Question   132
      D. EC External Relations in the Environmental Field   134
      E. The Existence and the Exclusivity Question
         in the Environmental Field   138
   3. Case Study on EC External Relations on Trade and
      the Environment 141
      A. Introduction on the Trade and Environment Issue   141
      B. The WTO Context and the Relevant GATT Provisions
         and Case Law   146
      C. The Application of Article XX(b) GATT   148
      D. The Application of Article XX(g) GATT   152
      E. The Case Law on the SPS Agreement   153
      F. The Case Law on the TBT Agreement   156
      G. Remarks on the EC's Behaviour in the Context of the
         WTO Case Law   157
   4. Case Study on EC External Relations on Climate Change   161

A. Introduction on the International Climate Change Regime   161
　　B. The EC and the International Climate Change Regime   162
　　C. Issues of International and EC Responsibility for
　　　　Non-compliance with the Climate Change Commitments
　　　　under the Kyoto Protocol   165
　5. Concluding Remarks   168

6. **A Climate of Change: An Analysis of Progress in EU and
   International Climate Change Policy**   171
　1. Introduction   171
　2. Climate Science: Spreading the Inconvenient Truth   173
　　A. Politics and IPCC's Fourth Assessment Report   173
　　B. Future Challenges for the IPCC   180
　3. Climate Change, Energy Security, and Competitiveness   183
　4. Mainstreaming Climate Change into the International
　　Policy Agenda   189
　　A. The G-8: from Gleneagles via Heiligendamm to Hokkaido   190
　　B. The Major Economies Meeting   192
　　C. Mainstreaming Climate Change in the UN System   194
　　D. Finance and Trade Ministers' Meetings in Bali   195
　　E. Mainstreaming: Challenges Ahead   197
　5. The Bali UN Climate Change Conference   198
　　A. A Heavy Agenda   198
　　B. A Difficult Final Spurt   200
　　C. The Bali Action Plan   201
　　D. A Shared Vision   202
　　E. Further Commitments for Developed Countries   204
　6. Actions by Developing Countries   205
　7. Conclusion: The Road from Bali   208

Index   213

List of Contributors

Jane Holder, Faculty of Laws, University College London, is a member of the Centre for Law and the Environment and the Centre for Law and Governance in Europe. Her area of expertise is environmental assessment, and her research interests include public participation in decision making processes, issues of locality, identity and solidarity, and communal responsibility for environmental protection. Recent publications include *Environmental Protection, Law and Policy* (with Maria Lee) (CUP, 2007), *Taking Stock of Environmental Assessment* (edited with Donald McGillivray) and *Emerging Commons* (edited with Tatiana Flessas, Special Issue of Social and Legal Studies).

Ingmar von Homeyer is a Senior Fellow at Ecologic, Institute for International and European Environmental Policy, Berlin and Brussels. He holds a PhD (2002) in Social and Political Sciences from the European University Institute, Florence, and studied political science at Free University Berlin. His work focuses mainly on EU environmental policy and governance. His recent publications include: 'Emerging Experimentalism in EU Environmental Governance', in Charles F. Sabel and Jonathan Zeitlin (eds.), *Experimentalist Governance in the European Union: Towards a New Architecture?* (OUP, forthcoming); 'The Role of the OMC in EU Environmental Policy: Innovative or Regressive?', *European Spatial Research and Policy*, 14 (2007) 1, 43–61; *Democracy in the European Union: Towards the Emergence of a Public Space* (edited with Liana Giorgi and Wayne Parsons) (Routledge, 2006); 'The EU Deliberate Release Directive: Environmental Precaution versus Trade and Product Regulation', in Sebastian Oberthür and Thomas Gehring (eds.), *Institutional Interaction in Global Environmental Governance. Synergy and Conflict among International and EU Policies* (The MIT Press, 2006), 259–283.

Maria Lee is professor of law at University College London where her teaching and research focuses on environmental law and policy. She taught previously at King's College, London. Her recent publications include *EU Environmental Law: Challenges, Change and Decision-Making* (Hart Publishing, 2005) and *Agricultural Biotechnology Regulation: Law, Decision-Making and New Technology* (Edward Elgar, 2008).

Jürgen Lefevere is Policy Coordinator, International Climate Change Negotiations, at the Environment Directorate-General of the European Commission in Brussels. He is representing the European Commission in high level climate change negotiations and is responsible for coordinating and defining the Commission's input in international climate change negotiations. Jürgen has been participating in the international climate change negotiations since 1998, initially as a legal advisor and negotiator for the Alliance of Small Island States (1998–2003). Jürgen has also played a key role in the design and implementation of the EU emissions trading system.

Massimiliano Montini is Associate Professor of European Union Law at the University of Siena (Italy), a fully qualified lawyer, member of the Italian Bar. He is also Director of the Environmental Legal Team (ELT), an University-based research and consultancy group located within the University of Siena. He is consultant to the Italian Ministry for the Environment, Land and Sea and is he is Deputy Director of the bi-annual Summer School in International and European Law, organised by the University of Siena, in co-operation with University College London.

Joanne Scott is Professor of European Law at University College London. She taught previously at Clare College, Cambridge and at the University of Kent and Queen Mary, University of London. She has been a visiting professor at Columbia, Georgetown and Harvard Law Schools, and a Jean Monnet Fellow at the EUI. She writes in the areas of EU law and WTO law, including the relationship between them. She is interested in exploring new approaches to governance in the EU, especially in environmental law. Her recent publications include a monograph on the SPS Agreement (OUP, 2007) and a co-edited (with Gráinne de Búrca) volume on Law and New Approaches to Governance in the EU and the US (Hart, 2007).

Table of Cases

Cassis de Dijon (Rewe-Zentrale)(Case 120/78) [1979] ECR 649 159
Commission v Council (Case 22/70) [1971] ECR 263 (AETR case) 131–4
Commission v Germany (Case C-476/98) [2002] ECR I-9855 132
Commission v Luxembourg (Case C-32/05) [2006] ECR I-11323 37
Commission v Italy (San Rocco) (Case C-365/97) [1999] ECR I-17773 37
Danish Bees (Case C-67/97) [1998] ECR I-8033 159
Danish Bottles (Case 302/86) [1988] ECR 4607 159
Drift-Nets (Case C-405/92) [1993] ECR I-6133 137
EC–Asbestos case, Report of the Panel, WT/DS/135/R 149–50, 156
EC–Biotech, Report of the Panel, WT/DS291/R 155
EC–Hormones case, Reports of the Panel, WT/DS36/R/USA and
    WT/DS26/R/CAN 154, 157
Hedley Lomas (Case 5/94) 68
Inter-Environment Wallonie (Case C-129/96) 83
Kramer (Cases 3, 4, 6/76) [1976] ECR 1279 . . . . 131, 137
Land Oberosterreich and Austria v Commission (Cases C-439/05 P and C-454/05 P),
    13 September 2007 73
Mangold (Case C-144/04) 83
Pfizer (Case T-11/99) 79
Preussen Elecktra (Case C-379/98) [2001] ECR I-2099 159
Retreaded Tyres (2007) 150
Shrimp/Turtle, Report of the Panel, WT/DS58/R 148, 150, 152–3
US Gasoline, Report of the Panel, WT/DS2/R 147–8, 150, 152, 156
Used Oils (Case 240/83) [1984] ECR 531 159
Walloon Waste (C-2/90) [1992] ECR I-4431 159

Table of Legislation

UK ACTS OF PARLIAMENT

2004 Planning and Compulsory Purchase Act 110

FOREIGN LEGISLATION

US Toxic Substances Control Act 79

EUROPEAN DIRECTIVES

Dir.67/548 on Dangerous Substances 62, 71, 72
Dir.75/440/EEC on Surface Water for Drinking Water Abstraction 10–11, 26
Dir.75/442/EEC on Waste 37
Dir.76/166/EEC on Bathing Water 10–11, 25
Dir.79/409/EEC on the Conservation of Wild Birds 169
Dir.80/778/EEC on Air Quality 10–11, 18
Dir.85/337/EC on EIA and Subsequent Modifications 123, 169
Dir.88/76/EC on Vehicles Emissions 12
Dir.88/609 EEC on Large Combustion Plants 13
Dir.90/313/EC on Public Access to Environmental Information 18
Dir.90/220 on the Release of Genetically Modified Organisms 155
Dir.91/156/EC on Waste 37
Dir.91/676/EEC on Urban Waste Water 21
Dir.92/43/EC on the Conservation of Natural Habitats and of Wild Fauna
  and Flora 34
Dir.96/42/EC on the Air Quality Framework 11, 18
Dir.96/61/EC on Integrated Pollution Prevention and Control 17–18, 169
Dir.96/63/EC on the Air Quality Framework 17–18
Dir.97/11/EC amending the EIA Directive 123
Dir.1999/30/EC on ambient air quality 11
Dir.1999/32/EC on Sulphur Content of Certain Liquid Fuels 169
Dir.2000/60/EC the Water Framework Directive 20–3, 25, 27–32, 35–8, 40–1,
                                                      44–5, 49–52, 55, 81, 83, 90
Dir.2001/18 amending the Directive on the Release of Genetically
  Modified Organisms 155
Dir.2001/42/EC on the Assessment of the Effects of Certain Plans on
  the Environment 37, 112
Dir.2001/80/EC on the Protection of Air from Large Combustion Plants 169
Dir.2001/2455/EC establishing the Priority Substances in Water Policy 32
Dir.2003/54/EC establishing Common Rules in Electricity 169
Dir.2003/55/EC establishing Common Rules in Natural Gas 169
Dir.2004/35 on Environmental Liability with regard to the Prevention and
  Remedying of Environmental Damage 42
Dir.2006/118/EC on Groundwater 26, 32

EUROPEAN REGULATIONS

Reg.4253/88 on the Structural Funds 95
Reg.258/97 on Novel Foods 155
Reg.1210/90/EEC Establishing the European Environment Agency 18
Reg.1049/2001 on Public Access to Documents 88
Reg.850/2004 on Persistent Organic Pollutants 81
Reg.1907/2006 REACH 56–61, 63, 65, 68–70, 73, 75, 78–82, 84–91

1

The Evolution of EU Environmental Governance

Ingmar von Homeyer

1. Introduction

Today the EU is a key driver of environmental policy in Europe and beyond. Not only is the 'environmental policy agenda of EU member states [...] now largely determined by the need to implement prevailing European law and to anticipate and shape European measures and action plans',[1] but the EU has also 'positioned itself as the world leader in the field of international environmental policy'.[2] At the heart of the EU's position lies the Union's environmental *acquis communautaire* which comprises over 200 major legal acts covering a comprehensive array of environmental issues.[3] This chapter argues that EU environmental policy developed over four major phases, each of which has given rise to a distinctive governance regime.[4] Although the early regimes have to some extent been modified as a result of the emergence of subsequent regimes, they nevertheless retain some of their core attributes. Consequently, EU environmental governance can be described as consisting of several interrelated but more or less distinct governance regimes.

---

[1] Jänicke, M. and Jörgens, H., 'New Approaches to Environmental Governance', in Jänicke, M. and Jacob, K. (eds.), *Environmental Governance in Global Perspective. New Approaches to Ecological and Political Modernisation* (Berlin: Freie Universität Berlin, 2006), p. 173.
[2] Kelemen, R. D., Globalizing EU Environmental Regulation, Paper prepared for a conference on Europe and the Management of Globalization, (Princeton University, 2007), p. 1.
[3] European Commission, How does a country join the EU? Latest developments published on the Internet at: <http://ec.europa.eu/enlargement/enlargement_process/accession_process/how_does_a_country_join_the_eu/negotiations_croatia_turkey/index_en.htm>.
[4] In contrast to concepts such as modes of governance (see, for example, de Búrca and Scott, *New Governance and Constitutionalism in the EU and US* (Hart Publishing, 2006) and Kohler-Koch and Eising, *The Transformation of Governance in Europe* (Routledge, 1999)), the term governance regimes is used here as a policy specific concept in the sense that a governance regime incorporates features and combinations thereof which are specific for a particular issue area. A particular legal base is a simple example.

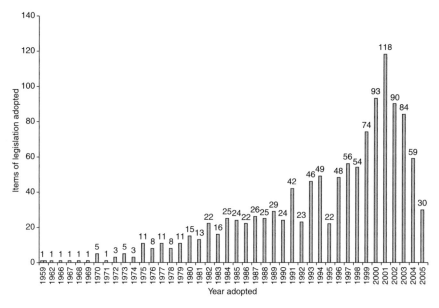

Figure 1. Number of items of EU legislation adopted each year 1967–2005
Source: Pallemaerts, M. (ed.), *Manual of Environmental Policy—The EU and Britain* (Maney, 2007), 2.1-2.

Environmental policy was not mentioned in the founding Treaties of the European Community (EC), but emerged in the early 1970s in the context of the Club of Rome's influential report on the limits of growth and the first major United Nations (UN) conference dealing specifically with the environment which was held in Stockholm in 1972. In the same year EC heads of state decided to establish a Community environmental policy and the Commission created the Environment and Consumer Protection Service (ECPS—the predecessor of the Commission's Directorate General (DG) for Environment).[5] In 1973 the Commission adopted the first EC Environmental Action Programme (EAP). However, until the Single European Act (SEA) came into force in 1987, EU environmental policy continued to develop in the absence of a distinct legal base in the Treaties. As illustrated in Figure 1, a rising number of legislative acts were nonetheless adopted in this period. This often happened on the basis of Article 235 TEC,[6] a gap filler clause which creates the possibility to adopt EU legislation in the absence of a more specific Treaty base. After a specific environmental legal base had been created in 1987, the number of legal acts adopted continued to rise significantly until 2001, but fell sharply thereafter.

---

[5] Jordan, A., 'Editorial introduction: the construction of a multi-level environmental governance system', *Environment and Planning C: Government and Policy*, 17 (1999) 4.
[6] Now Article 308 TEC.

Table 1. Overview of EU environmental governance regimes

Governance regime	Environment	Internal Market	Integration	Sustainable Dev.
Year of emergence	1972	1982	1992	1998

Table 1 provides an overview of the four major stages in which EU environmental governance developed and of the corresponding EU environmental governance regimes: the 'environment regime' which dates back to 1972; the 'internal market regime' (1982); the 'integration regime' (1992); and the 'sustainable development regime' (1998). The identification of the four regimes and their characteristics is largely based on empirical observation. However, it should be noted that the gradual evolution and subsequent coexistence of the regimes seems to correspond well to the notion of institutional layering developed by historical institutionalist scholars such as Kathleen Thelen. Layering occurs if actors create alternative institutions around existing institutions. Depending on the relative growth rates and overlap between existing and alternative institutions, the latter may eventually replace the former, thereby creating a new path of institutional development. Alternatively, original and alternative institutions may coexist for prolonged periods of time and interact in various ways.[7] Such coexistence seems to apply to EU environmental policy where new governance regimes have modified, but not replaced pre-existing ones.

The main purpose of this chapter is to identify and describe the four EU environmental governance regimes. Therefore no attempt beyond the reference to institutional layering is made to offer a systematic analysis of the factors which preserved important aspects of the early regimes in the face of the emergence of newer regimes. Similarly, although the factors which led to emergence of the various regimes in the first place are often partly reflected in the regime characteristics—for example in the respective problem-solving focuses—only a limited attempt to explain the emergence of the regimes will be made.

This chapter proceeds as follows: the next section presents a general overview of some of the most important factors affecting the evolution of EU environmental governance. Following this overview, the remainder of this chapter describes each of the EU environmental governance regimes in terms of the main formal or informal 'legal base' of measures associated with the regime, the prevailing decision-making mode, underlying political dynamic, and type of political justifications, objectives, and instruments. In addition, key factors contributing to the emergence of each governance regime as well as examples of measures

---

[7] Streeck, W. and Thelen, K., 'Introduction: Institutional Change in Advanced Political Economies', in Streeck, W. and Thelen, K. (eds.), *Beyond Continuity: Institutional Change in Advanced Political Economies*, (Oxford: Oxford University Press, 2005), pp. 33–34.

associated with the regimes are identified. The conclusion discusses the relevance of the findings with respect to continuity and change of EU environmental governance and presents potential implications for its future evolution.

2. Factors Affecting the Evolution of EU Environmental Governance

The factors which affected EU environmental governance to some extent varied over time both in terms of their qualitative characteristics and their relative importance. Nonetheless an examination of the evolution of EU environmental governance suggests that four sets of factors have generally been relevant: environmental and economic conditions; international political developments and commitments; major changes of the EU polity; and the interests and ideological orientations of certain actors.

A. Environmental and Economic Conditions

The perceived nature of environmental and economic trends has often affected EU environmental governance. As illustrated further below, environmental problems were frequently perceived as acute threats to health or the environment throughout much of the 1970s and 1980s. These perceptions tended to favour hierarchical, top-down environmental governance at national and EU levels. Similarly, globalization and increasing economic competition in the 1990s and beyond contributed to the turn towards more deliberative and flexible environmental governance patterns which aim to integrate economic considerations into environmental policy-making.[8]

B. International Political Developments and Commitments

Regional, and in particular global political developments and commitments have also affected EU environmental governance.[9] As mentioned above, the UN Stockholm conference in 1972 contributed to the creation of the first EU environmental governance regime. Other examples include the 1992 UN Earth Summit and the Kyoto Protocol or, at the regional level, the Aarhus Convention which contributed to the adoption of improved public information and participation procedures in the EU.[10]

---

[8] Cf. Jänicke and Jörgens, n. 1; Holzinger, K. and Knill, C., *European Environmental Governance in Transition?* (Bonn: Max-Planck-Projektgruppe Recht der Gemeinschaftsgüter, 2002).
[9] Oberthür, S., and Gehring, T., *Institutional Interaction in Global Environmental Governance. Synergy and Conflict among International and EU Policies* (Cambridge/London: The MIT Press, 2006).
[10] Cf. Pallemaerts, M., Wilkinson, D., Bowyer, C., Brown, J., Farmer, A., Farmer, M., Herodes, M., Hjerp, P., Miller, C., Monkhouse, C., Skinner, I., ten Brink, P. and Adelle, C., *Drowning in Process? The Implementation of the EU's 6th Environmental Action Programme Report for the European*

C. Major Changes of the EU Polity

Two types of major changes of the EU polity frequently had an impact on EU environmental governance. The first type concerns major Treaty changes. Examples include the Single European Act which, as mentioned above, created a specific Treaty base for EU environmental policy and provided for various other environmental clauses. In addition the SEA introduced qualified majority voting in the Council of Ministers for Internal Market related EU environmental legislation. Subsequent Treaty changes extended both the EU's environmental competences and qualified majority voting, introduced certain environmental principles into the Treaties etc.[11]

EU enlargement constitutes a second type of major change of the EU polity affecting EU environmental governance.[12] Albeit with a certain delay, the accession of Greece, Portugal, and Spain in the 1980s significantly contributed to focusing EU environmental governance more strongly on implementation in the 1990s[13]—a trend which was further reinforced at the time by the prospect of Eastern enlargement. For its part, Northern enlargement (Austria, Finland, Sweden) in the 1990s reinforced efforts to promote sustainable development and the integration of environmental concerns into other policies.[14]

Interests and ideological orientations

The interests and ideological orientations of a number of actors appear to have had a strong impact on EU environmental governance. This concerns, in particular, the interests of certain Member State governments, the ideological orientation of the Commission and of the European Parliament, as well as the interests of non-state actors, in particular business associations and environmental NGOs. For instance, as illustrated further below, groups of Member States which acted as environmental pace-setters[15] contributed to the formation of different EU environmental governance regimes. The shift towards a more conservative

---

[11] Jordan, n. 5, pp. 11–12.

[12] Cf. von Homeyer, I., 'Differential Effects of Enlargement on EU Environmental Governance', in Carmin, J. and VanDeveer, S. D. (eds.), 'EU Enlargement and the Environment. Institutional Change and Environmental Policy in Central and Eastern Europe', *Environmental Politics*, Special Issue, 13 (2004) 52–76.

[13] Cf. Börzel, T.A., *Environmental Leaders and Laggards in Europe: Why there is (Not) a Southern Problem* (London: Ashgate, 2003).

[14] von Homeyer, I., 'Emerging Experimentalism in EU Environmental Governance', in Sabel C. F. and Zeitlin, J. (eds.), *Experimentalist Governance in the European Union: Towards a New Architecture?* (Oxford University Press, forthcoming); Lenschow, A., 'Greening the European Union: An introduction', in Lenschow, A., *Environmental Policy Integration. Greening sectoral policies in Europe* (London: Earthscan, 2002), p. 11.

[15] Andersen, M. S. and Liefferink, D., *European Environmental Policy: The Pioneers* (Manchester: Manchester University Press, 1997); Héritier, A., Knill, C., and Mingers, S., *Ringing the Changes in Europe. Regulatory Competition and the Redefinition of the State: Britain, France, Germany*, (Berlin, New York: De Gruyter, 1996).

Table 2. Key features of EU environmental governance regimes

	Environment	Internal Market	Integration	Sustainable Development
Period	1972	1982	1992	1998
Aim	Acute problems	Completion of Internal Market	Efficiency-effectiveness	Persistent problems
Legal base	Art. 235 (308)	Art. 100a (95)	Art. 130s (175)	Art. 6
Decision-making	Technocratic	Voting	Negotiation	Comparison, mutual learning
Political dynamic	Politicization	Economic competitiveness	Implementation	Destabilization
Instrument	Regulation	Harmonization	'Integrating' framework directives, 'horizontal legislation'	'Reflexive' framework directives, long-term strategies
Type of objectives	Quality objectives	Emission limits	Broad, integrative objectives	EU-wide long-term targets
Justification	scientific	technological	pragmatic	normative
Examples	Bathing Water Directive, Drinking Water Abstraction Directive, Air Quality Directive (1980)	Vehicle Emissions Directives, (Large Combustion Plants Directive)	IPPC Directive, Air Quality Framework Directive	EU-Sustainable Development Strategy, Some Thematic Strategies and corresponding legislation, Water Framework Directive

European Commission and Parliament in 2004 provides an example of the impact of ideology on EU environmental governance. Among other things, this shift appears to have further weakened the political influence of the long-standing de facto alliance between the Environment Council, the Environment Committee, and the Commission's DG Environment.[16]

3. EU Environmental Governance Regimes

As illustrated in Table 2, important features distinguishing the four EU environmental governance regimes include their problem-solving focus, their legal base, decision-making mode, underlying political dynamic, mode of justification, as well as types of objectives and instruments used. More specifically, the *problem-solving focus* of a given environmental governance regime refers to the major issues which measures adopted under the regime are meant to address. Examples include direct threats to human health and the environment resulting from pollution, increasing the effectiveness of environmental measures or reducing their effects on economic competitiveness. While the *legal base* which is most characteristic for a regime may often refer to the predominant formal legal Treaty base for measures associated with the regime, it can also refer to more abstract Treaty provisions which provide a more general rationale for these measures, such as the requirement in Article 6 TEC to integrate environmental concerns into all Community policies ('Environmental Policy Integration'). Similarly, the typical *decision-making mode* does not simply coincide with predominant formal decision-making rules associated with a regime, such as qualified majority voting in the Council and the Co-decision Procedure, but also includes informal rules which govern the application of the formal rules. *Political dynamic* refers to the prevailing type of incentives, such as politicization of an issue or a particular substantive concern, which propel decision-makers into action. The *mode of justification* denotes the dominant type of discourse used or referred to by decision-makers to explain and defend measures associated with a particular regime. Scientific or more pragmatic arguments relating, for example, to the effectiveness or efficiency of a measure are a case in point. The *type of objective* refers to characteristics such as specificity and substantive focus of the objectives of measures. For instance, objectives may typically have the form of emission limits under one regime, whereas under a different regime environmental quality objectives may be more prominent. Finally, the preferred *type of instrument* varies among environmental governance regimes. Whereas detailed EU directives stand

---

[16] Jordan, A., 'The politics of a multi-level environmental governance system: European Union environmental policy at 25', *CSERGE Working Paper*, PA 98-01 (Norwich: University of East Anglia, 1998), pp. 13, 16; Pallemaerts et al., *Drowning in Process? The Implementation of the EU's 6th Environmental Action Programme Report for the European Environmental Bureau* (London: IEEP, 2006).

out under one regime, flexible framework legislation or economic instruments may be the instruments of choice for another regime.

The following sections offer a description of the characteristics of the four governance regimes. However, it should be pointed out that the various features usually emerged over a period of time and that some characteristics emerged earlier than others. It is not always clear when exactly a particular governance regime emerged. The years in the Table are therefore indicative only.[17]

A. The Environment Regime

The original European environmental governance regime dates back to the early 1970s. At the time, the main challenge of environmental policy was frequently perceived to be related to acute threats to human health and the environment.[18] The first EU Environmental Action Programme, which was adopted in 1973, exemplifies this focus. Although the Programme clearly recognized that environmental protection extends beyond addressing acute threats, those threats formed the main focus of the programme. This is reflected, for example, in the short, two-year timeframe for the activities which were foreseen as part of the programme. In addition, a large majority of these activities concerned the reduction of pollution and nuisances rather than improving the environment more generally. Within the emphasis on pollution reduction, the focus was on substances which posed health risks, such as toxic substances, in particular as components of waste and contributing to water pollution.[19]

As mentioned above, before 1987 legislative measures to protect the environment frequently had to be based on Article 235 TEC. This general gap filler clause enabled the adoption of European legislation if 'action by the Community should prove necessary to attain ... one of the objectives of the Community and this Treaty has not provided the necessary powers'. As was generally the case for the adoption of legislation, Article 235 required unanimous agreement by the Member States represented in the Council. But this formal requirement was 'softened' by the technocratic decision-making mode which characterizes the

---

[17] The various phases in which the governance regimes emerged to some extent correspond to Christian Hey's analysis of EU environmental strategies, in particular the Environmental Action Programmes. See Hey, C., 'EU Environmental Policies: A Short History of the Polic Strategies', in Scheuer, S. (ed.), EU Environmental Policy Handbook (Brussels: European Environmental Bureau, 2005). However, the discussion in this chapter focuses on governance rather than programmatic shifts. This may explain why some of the programmatic shifts identified by Hey seem to correspond to governance regimes which, however, did not emerge until several years later.

[18] K. Jacob et al., 'Einleitung Politik und Umwelt-Modernisierng politischer Systeme und Herausforderung an die Politikweissenschaft' in Jacob, K. et al. (eds.), *Politik und Umwelt*, Polititische Vierteljahresschrift, Sonderheft 39 (Wiesbaden: VS Verlag, 2007), p. 13.

[19] Cf. Council of the European Communities, 'Declaration of the Council of the European Communities and of the representatives of the Governments of the Member States meeting in the Council on the programme of action of the European Communities on the environment', [1973] OJ C112, 01–02.

environment regime and which tended to prevail in particular in the European Commission at the time.[20] For example, the first EAP assumed that expertise and scientific knowledge provided the main basis—as well as the most important justification—for taking environmental decisions at EU level.[21] Consequently, the Programme called for the creation of a 'Consultative Committee on Environment *Research*' [emphasis added] composed of Member State representatives as the body to assist the Commission in carrying out the Programme. Frequently, the 'Commission was left to work up proposals relatively unsupervised' as 'political leaders . . . seemed content to let environmental policy develop under its own impetus'[22] and powerful Commission services, such as those dealing with industry or trade, 'saw very little point in scrutinising every proposal issued by the ECPS',[23] the Commission's small environmental service at the time.

However, the coexistence of the strong technocratic orientation of the Commission with the unanimity rule in the Council occasionally produced significant tensions if the Commission's proposals touched on strong national interests. In these cases the proposals were radically watered down by one or more Member States. For example, this was the fate of the Commission's first proposal for a directive addressing safety issues related to genetically modified organisms. Having encountered strong British resistance, the 1976 Commission proposal, which had been justified mainly in scientific terms, ended up as a legally non-binding recommendation after four years of controversy.[24] Similar tensions between the Commission's technocratic approach and unanimous decision-making in the Council hampered the adoption of EU legislation on the emission of dangerous substances into water[25] and may explain the fact that EU environmental measures adopted in the period of EU environmental policy-making dominated by the environment regime tended to grant Member States more implementation flexibility than measures adopted in subsequent years.[26]

Interestingly, the rise of the environment regime and its technocratic decision-making mode coincided with a certain politicization of environmental policy in

---

[20] Jordan, n. 5, p. 5.
[21] Council of the European Communities, n. 19; As Shapiro describes this 'softening': 'Turning over key decisions to technical experts with one or more drawn from each Member State can provide the appearance of national representation while in reality achieving only the representation of the interests technical experts derive from their particular expertise'. See Shapiro, M., ' "Deliberative", "Independent", Technocracy v. Democratic Politics: will the Globe Echo the EU?' (2005) 68 *Law and Contemporary Problems* 354.
[22] Jordan, n. 5, p. 4.
[23] Ibid., p. 5.
[24] Von Homeyer, I., *Institutional Change and Governance in the European Union: The Case of Biotechnology Regulation* (PhD Thesis, EUI, 2002), pp. 73–97; Cantley, M. F., 'The Regulation of Modern Biotechnology: A Historical and European Perspective', in Rehm, H. J. and Reed, G. (eds.), *Biotechnology Vol. 12, Legal, Economic and Ethical Dimensions*, (Weinheim: VCH, 1995), pp. 519–23.
[25] Jordan, n. 5, p. 7.
[26] Holzinger, K. Knill, C., and Schäfer, A., *European Environmental Governance in Transition?* (Bonn: Max-Planck-Projektgruppe Recht der Gemeinschaftsgüter, 2002), pp. 20–2.

some Member States, in particular Germany, where the governing Social Democrats (SPD) and Liberals (FDP) used environmental issues to sharpen their political profile in the early 1970s.[27] Germany and several other countries such as Denmark, the Netherlands and, occasionally, France, therefore provided some backing for EU environmental policy against more reserved Member States, in particular the UK.[28] Governance patterns resembling the decision-making mode of the environment regime, including the coexistence of a technocratic decision-making mode and a certain degree of politicization, can still be observed in certain more recent instances of EU environmental policy-making, for example regarding genetically modified organisms.[29]

The environment regime's combination of a focus on acute health and environmental threats with a technocratic decision-making mode implies a preference for a regulatory approach based on environmental quality objectives. This is because quality objectives—provided they are grounded in adequate scientific knowledge and are properly implemented—can ensure that a predetermined state of the environment characterized by the absence of acute threats will indeed be attained. In addition, quality objectives can easily be combined with a technocratic decision-making mode which privileges expert input into the decision-making process.

In terms of instruments, the environment regime typically relies on legally binding, top-down regulation. The regime's focus on acute health and environmental problems and its reliance on expertise in the decision-making process partly account for this. More specifically, the real or perceived acuteness of problems and the science-based substantive contents of legislation require quick and full implementation. Legally binding, top-down regulation allows for exact instructions to implementing bodies, is associated with a high level of obligation, and enables the use of mostly pre-established administrative structures. On the down side, this science-based, top-down regulatory approach tends to be costly, among other things due to low flexibility and relatively high administrative costs.

The 1975 Directive on Surface Water for Drinking Water Abstraction,[30] the 1976 Bathing Water Directive,[31] and the 1980 Air Quality Directive[32] exemplify

---

[27] Cf. Wurzel, R. 'Germany', in Jordan, A. and Lenschow A. (eds.), *Innovation in Environmental Policy: Integrating the Environment for Sustainability* (Cheltenham: Edward Elgar, forthcoming).

[28] As global economic conditions deteriorated in the second half of the 1970s, German social democratic and liberal support for environmental measures weakened. However, this was partly compensated by growing grass roots pressure and the birth of the Green Party in Germany.

[29] Cf. Dabrowska, P., 'EU Governance of GMOs: Political Struggles and Experimentalist Solutions?', in Sabel, C. F. and Zeitlin, J. (eds.), *Experimentalist Governance in the European Union: Towards a New Architecture?* (Oxford: Oxford University Press, forthcoming); von Homeyer, I., 'Participatory Governance in the EU', in Giorgi, L., von Homeyer, I., and Parsons, W. (eds.), *Democracy in the European Union. Towards the Emergence of a Public Sphere*, (Routledge: London, 2006).

[30] Directive 75/440/EEC.
[31] Directive 76/166/EEC.
[32] Directive 80/778/EEC.

regulatory measures associated with the environment regime. The Drinking Water Abstraction and Bathing Water Directives are very early pieces of EU environmental legislation. Aimed primarily at ensuring public health, the Directives fixed parameters with which surface water used for the abstraction of drinking water and bathing water had to comply. Despite implementation problems, the ambitious Bathing Water Directive contributed significantly to improvements in bathing water quality, whereas the Drinking Water Abstraction Directive had more limited effects, partly because of its moderate level of ambition. After 30 years the Bathing Water Directive was replaced in 2006 by a revised Directive which, however, maintains the basic approach of the original Directive. The Drinking Water Abstraction Directive was repealed in 2007 when its functions were integrated into the Water Framework Directive.

The 1980 Air Quality Directive sets air quality standards for sulphur dioxide and particulate matter. Once again, the primary aim is 'to protect human health in particular'.[33] The Directive refers to the scientific data underpinning the work of the World Health Organisation (WHO) to justify its approach. While the Directive's standards appear to have been too weak to produce significant effects in some Member States, there was a stronger impact in more polluted areas.[34] Partly in response to new scientific knowledge, in particular regarding the health impact of fine particulate matter, the Directive was replaced by a daughter directive[35] to the 1996 Air Quality Framework Directive[36] in two steps in 2001 and 2005.

B. The Internal Market Regime

From around 1982 the completion of the Community's internal market rose to the very top of the European agenda, culminating in the adoption of the Single European Act (SEA) in 1986 and providing the European integration process with renewed momentum.[37] European environmental governance reflected these developments in that harmonization to avoid trade barriers and the creation of a level playing field for companies competing within the internal market became a dominant concern behind the pursuit of environmental policy,[38] which was supported by a fully fledged Directorate General (DG Environment) from 1981 on.

---

[33] Preamble, Directive 80/778/EEC.
[34] Cf. Pallemaerts, M. (ed.), *Manual of European Environmental Law: The EU and Britain* (Maney, 2007), Section 6.4–6, published on the Internet at: <http://www.ieep.eu/publications/manual.php>; Hey, n. 17, III.2.
[35] Directive 1999/30/EC.   [36] Directive 96/42/EC.
[37] Young, A. and Wallace, H., 'The Single Market: A New Approach to Policy', in Wallace, W. and Wallace, H. (eds.), *Policy making in the EU*, 4th edn. (Oxford: Oxford University Press, 2000), p. 93.
[38] Hey, n. 17, at III.3.

Once the SEA had entered into force in 1987, environmental measures no longer needed to be based on Article 235 TEC because the SEA had introduced a specific legal base—Article 130s TEC—for Community environmental legislation. At the time, however, the provisions contained in the equally new Article 100a TEC on the approximation of laws to complete the internal market were of even larger environmental significance.[39] First, Article 100a TEC required legislative proposals to conform to a high level of environmental protection and allowed Member States to maintain or adopt standards that exceeded Community norms under certain conditions. Second, the Article was subsequently frequently used as a legal base for much European legislation. This reflected the political momentum of the internal market project and the fact that—unlike other provisions, including Article 130s TEC—Article 100a TEC allowed for qualified majority voting in the Council.[40]

Reflecting these developments, the internal market regime is characterized by a decision-making mode in which behaviour induced by majority voting rules—building qualified majorities or blocking minorities—is a major element. The adoption of the 1988 Vehicles Emissions Directive[41] illustrates this point: several Member State governments which had initially blocked the legislative process with veto threats were outvoted in the Council after the SEA had abolished the national veto for internal market related measures. Broadly similar coalitional dynamics—this time also including the newly empowered European Parliament—characterized the adoption of a related Directive on emissions from small cars a year later.[42]

The political incentives driving EU environmental policy under the internal market regime centre on economic competition. For Member States which, for political or other reasons, have adopted—or plan to adopt—relatively strict environmental standards applying to the production process ('process standards') it often makes economic sense to try to 'upload' these standards to the EU level. By making such standards binding on all EU Member States, environmentally progressive countries can avoid putting their industries at an economic disadvantage vis-á-vis their European competitors.[43] Similarly, there can also be an economic rationale for uploading of existing or planned national environmental standards applying to products ('product standards') if such standards increase the

---

[39] Cf. Council of the European Communities, 'Resolution of the Council of the European Communities and of the representatives of the Governments of the Member States, meeting within the Council on the continuation and implementation of a European Community policy and action programme on the environment (1987–1992)', [1987] OJC 328, 01–44.

[40] Jordan, n. 5, p. 9.

[41] Directive 88/76/EC.

[42] Boehmer-Christiansen, S. and Weidner H., *The Politics of Reducing Vehicle Emissions in Britain and Germany* (London: Cassell, 1995); K. Holzinger, *Politik des kleinsten gemeinsamen Nenners? Umweltpolitische Entscheidungsprozesse in der EG am Beispiel der Einführung des Katalysators* (Berlin: Edition Sigma, 1994).

[43] Héritier, A., Knill, C., and Mingers, S., *Ringing the Changes in Europe. Regulatory Competition and the Redefinition of the State: Britain, France, Germany* (Berlin, New York: De Gruyter, 1996).

price of exports, for example as a consequence of strong economies of scale which render the introduction of different product lines for different national markets inefficient. More importantly, however, the European Commission and Member States with low product standards are likely to be concerned that environmental requirements in progressive countries could undermine the internal market if such countries were to ban imports of products not complying with their high standards. Given that Article 100a TEC requires a high level of environmental protection and under certain circumstances allows countries to maintain higher standards than those agreed at the European level, there are significant incentives for Member State governments and the Commission to adopt relatively high common product standards at EU level in an effort to avoid trade barriers.[44]

Because considerations of economic competitiveness are key to the internal market regime, emission limits are the preferred type of objective. Common emission limits ensure that producers and products will have to comply with the same requirements independently of their geographical location within the internal market. In this way, emission limits provide for a level playing field and the removal of trade barriers.

Harmonization measures feature as the type of instrument that is most closely associated with the internal market regime. As with the environment regime, these measures frequently take the form of legally binding, top-down regulation. However, the administrative and budgetary burden is likely to be lower than in the case of the environment regime because ensuring compliance largely falls to private actors and, in the case of product standards, even tends to be in the economic interest of at least some producers, namely those with significant export markets in highly regulated Member States which could ban imports in case of non-compliance.

Given that the internal market regime is primarily concerned with competition within the internal market rather than with environmental outcomes, the justification of measures relies less on science and more on arguments about technical feasibility of pollution abatement measures and, to a lesser extent, economic implications, in particular for producers in different Member States. For example, in the case of the vehicles emission directives, the availability of catalytic converters capable of significantly reducing car emissions provided a crucial argument in favour of stricter emission limits. More generally, concepts of technical feasibility, in particular the German concept of best available technology (BAT), tended to prevail even in instances of decision-making where the national veto remained viable. This was the case, for example, with the Large Combustion Plants Directive.[45]

---

[44] von Homeyer, n. 12, pp. 52–76; Scharpf, F. W., 'Community and Autonomy: Multilevel Policymaking in the European Union', *Journal of European Public Policy*, 1 (1994), 219–39.

[45] Directive 88/609/EEC; Héritier, n. 43 and Héritier, A. (1995), *Die Koordination von Interessenvielfalt im europäischen Entscheidungsprozeß und deren Ergebnis: Regulative Politik als "Patchwork"*, MPIFG Discussion Paper 95/4, Max-Planck-Institut für Gesellschaftsforschung, Cologne, pp. 23–4.

The vehicles emissions directives also exemplify other aspects of the internal market regime, in particular the relevance of considerations of economic competition. Germany provided the main political impetus for the adoption of the directives. The German government was under strong pressure to improve air quality because of widespread public concern over forests suffering ('dying') from the effects of air pollution. In addition, the Green party was significantly increasing its appeal among voters. Against the background of threats by the German government to unilaterally introduce stricter car emission limits, the Commission proposed common European standards which, however, were strongly opposed by the governments of France and Italy. French and Italian producers of small cars argued that for technical and economic reasons the proposed emission limits would distort competition with the German manufacturers of larger cars. By contrast, the supporters of higher standards justified their position, among other things by referring to the example of California, where strict emission limits had already been adopted.

C. The Integration Regime

The integration regime focuses mainly on issues of efficiency and effectiveness of EU environmental measures. Several factors contributed to the emergence of the regime in the early 1990s, including the rising number of EU measures which had been adopted in particular towards the end of the internal market regime and which needed to be implemented in the following years. As Jordan[46] points out, between 1989 and 1991 'the Environment Council adopted more environmental policies than it had in the previous 20 years'. Following this burst of legislative activity the focus of EU environmental policy-making shifted to implementation, in particular the costs and effectiveness of the measures which had been adopted. As illustrated further below, this shift was reinforced by several other factors, in particular EU enlargement, the rise of the sustainable development agenda, issues of EU legitimacy and subsidiarity, as well as changing priorities in EU Member States.

The Maastricht Treaty, which had been agreed in late 1991 and came into force almost two years later, modified the environmental Treaty provisions which had first been introduced by the SEA. In particular, the Treaty chapter on the environment was revised. The new Article 175 TEC (former Article 130s TEC) extended QMV and significantly enhanced the decision-making role of the European Parliament in many, though not all, areas of EU environmental policy. The fact that the national veto had been dropped for many issue areas was one reason why Article 175 TEC became the main legal base for the integration

---

[46] Jordan, A., 'The politics of a multi-level environmental governance system: European Union environmental policy at 25', *CSERGE Working Paper*, PA 98–01 (Norwich: University of East Anglia, 1998), p. 11.

regime. More importantly, however, Article 175 TEC allowed for a shift away from the regulatory harmonization approach of the internal market regime towards the integration regime's focus on economic efficiency and environmental effectiveness—both of which frequently require a certain degree of flexibility and decentralization at the cost of harmonization.

Despite the availability of QMV, the integration regime's decision-making mode tends to be shaped less by voting rules in the Council than by extensive negotiations among Member States, with the European Parliament under the Co-decision Procedure, and with a broad range of other state and non-state stakeholders who became increasingly involved in EU environmental policy-making during the 1990s. Several factors, including a rising concern with economic competitiveness at the global level, may explain this shift towards more inclusive, networked governance.[47] Perhaps most importantly, the integration regime's focus on effectiveness and efficiency contributed to the emergence of more inclusive policy networks because it drew attention to the variation in national and regional conditions which influence the effectiveness and efficiency of environmental measures. Relevant differences concern, among other things, ecological and economic conditions, administrative capacities and styles as well as variation in the perceptions of costs and benefits. Rather than harmonization (internal market regime) or highly aggregated scientific knowledge (environment regime), broad participation by stakeholders and experts is required to identify, and agree on, locally adapted, flexible solutions. Under these circumstances voting and aggregated scientific knowledge lose part of their legitimacy and practical value in the decision-making process.

But there is a second reason why the adoption of differentiated, flexible regulation tends to increase the need to consult widely. Regulatory flexibility means that the benefits of regulating at the EU level in terms of the internal market tend to be relatively small because national regulations are only partly harmonized. Therefore these benefits can no longer be used to compensate actors who would otherwise oppose EU measures.[48] Other, often more direct ways of taking the concerns of potential 'losers' into account must be identified and agreed. Once again, this tends to require intensive consultation of the actors concerned.

Reflecting the emphasis on efficiency and effectiveness and an increasingly inclusive, negotiation-based decision-making mode, the integration regime justifies EU environmental measures mainly in pragmatic and procedural terms. The way in which the 5th EAP, which was adopted in 1992, introduces and employs the concept of 'shared responsibility' to 'translate' the Maastricht Treaty's Subsidiarity Principle into 'operational terms' illustrates this. According to the

---

[47] Holzinger, n. 26; Lenschow, A., 'Transformation in European Environmental Governance', in Kohler-Koch, B. and Eising, R. (eds.), *The Transformation of Governance in Europe* (London: Routledge, 1999).
[48] Cf. Majone, G., *Regulating Europe* (London: Routledge, 1996).

5th EAP, shared responsibility 'involves not so much a choice of action at one level to the exclusion of others but, rather, a mixing of actors and instruments at the appropriate levels'.[49] This implies that the appropriate level of action is neither externally given nor predetermined, but arises pragmatically; because 'sustainable development can only be achieved by concerted action on the part of all the relevant actors working together in partnership',[50] a critical criterion for determining effectiveness and efficiency is procedural, focusing on whether all relevant actors have been included in decision-making and implementation.

Efforts among Member States to 'up-load' their particular national regulatory approaches to the EU level provide the main political incentives for the pursuit of EU environmental policy under the integration regime. Among other things, successful up-loading allows Member States to stick to, and further entrench, their preferred regulatory approaches in the face of EU measures.[51]

At the same time, a demand for new instruments arose from several developments: on the one hand, it became increasingly clear in the early 1990s that the effectiveness of existing EU environmental measures was limited in that, despite fewer acute environmental problems, overall environmental trends were still deteriorating.[52] This was partly attributed to a lack of integration of environmental measures which led to inconsistencies and 'problem transfer' in the sense that certain measures merely solved one environmental problem at the cost of creating, or contributing to, a different one.

On the other hand, costs and administrative aspects came to the forefront. In the process of implementing the host of EU environmental measures adopted in the late 1980s and early 1990s, Member State governments' awareness of the associated financial and administrative burden had increased. In addition, the implications of EU enlargement for EU environmental policy became increasingly clear as Greece, Portugal, and Spain—the three southern European Member States which had joined the EU in the 1980s—failed to effectively implement important pieces of EU environmental legislation, at least partly due to lack of financial and administrative means. The prospect of many Central and Eastern European countries joining the EU further highlighted the significance of the 'implementation deficit' because these countries faced similar constraints. Moreover, the 1992 Rio Earth Summit and, more generally, the rise of the sustainable development paradigm emphasized the link between economic development and costs on the one hand and environmental protection on the

---

[49] Council of the European Communities, 'Resolution of the Council and the Representatives of the Governments of the Member States, meeting within the Council on a Community programme of policy and action in relation to the environment and sustainable development—A European Community programme of policy and action in relation to the environment and sustainable development', [1993] OJ C138/5 01–93 at 78.
[50] Ibid.
[51] Börzel, T. A., 'Shaping and Taking EU Policies: Member State Responses to Europeanization', Queen's Papers on Europeanisation, 2 (Belfast: Queen's University, 2003), p. 4; Héritier, n. 43.
[52] Council of the European Communities, n. 49.

other. Together, these trends created a demand for new environmental solutions and instruments which allowed for more efficient and effective implementation, for example flexible, integrated regulation, economic instruments, and information-based approaches.[53]

The UK, which had mostly tried to obstruct EU environmental policy in the 1970s and 80s, was particularly well positioned to meet this demand.[54] Spurred on by the increase of the Green vote in the late 1980s, the British government took a more constructive and proactive role in some areas of environmental policy, including at the EU level, while simultaneously calling for a re-nationalization of other EU environmental measures in the name of subsidiarity. In contrast to Germany's regulatory approach, which had strongly influenced EU environmental policy in the 1980s, important aspects of the British approach, in particular the concept of Integrated Pollution Control (IPC) as set out in the 1990 British Environmental Protection Act, were largely untested at the EU level and seemed suited to address the demand for more flexible, integrated regulation. While the UK became more involved in EU environmental policy, the opposite applied to Germany, which was preoccupied with the economic repercussions of unification.

The integration regime's environmental measures are characterized by vague objectives which leave room for flexibility and an integrated approach. The 1996 Integrated Pollution Prevention and Control (IPPC) Directive[55] and the Air Quality Framework Directive[56] provide examples. The IPPC Directive covers and 'integrates' the environmental media air, water, and land and aims 'to achieve a high level of protection of the environment taken *as a whole*'.[57] Similarly, the Air Quality Framework Directive aims 'to avoid, prevent or reduce harmful effects on human health and the environment as a whole' and 'to maintain ambient air quality where it is good and improve it in other cases'.[58]

Framework Directives are the most characteristic regulatory instruments associated with the integration regime. Reflecting their integrative ambition, these directives tend to have a relatively broad scope. They can cover various environmental media, quality and emission limits, or a range of pollutants that were previously regulated separately. The framework directives establish binding procedures, for example regarding planning, permitting, measurements, and reporting, or regulate the relationship among related and 'daughter' directives which, in turn, contain the relevant substantive requirements. This procedural approach is complemented by 'horizontal' measures which help to render the procedures effective,

---

[53] Cf. Von Homeyer, n. 12, pp. 52–76; Jordan, A., Wurzel, R., and Zito, A., *New Instruments of Environmental Governance: National experiences and prospects*, (Taylor and Francis Group: London, 2003).

[54] Héritier, A., Knill, C., and Mingers, S., *Ringing the Changes in Europe. Regulatory Competition and the Redefinition of the State: Britain, France, Germany* (Berlin, New York: De Gruyter, 1996), pp. 24–6.

[55] Directive 96/61/EC.
[56] Directive 96/63/EC.
[57] Directive n. 55, Article 1, emphasis added.
[58] Directive, n. 56, Article 1.

for example by enhancing the availability of information and by bestowing rights on stakeholders. For instance, in 1990 the Community adopted its first Directive on public access to environmental information.[59] Other measures include legislation to create the European Environment Agency (EEA) to provide the Commission and the Member States with environmental information[60] and a programme to financially support EU-level environmental NGOs.[61]

There are several other aspects of the IPPC and Air Quality Framework Directives typical of measures associated with the integration regime. For instance, the Treaty's specific environmental provisions serve as the legal base for the directives and, as mentioned above, the directives follow an integrative approach, but their substantive objectives remain vague. Applicable limit values and similar instruments are mostly contained in daughter directives or related legislation, while the two Directives themselves provide detailed procedural requirements, for example concerning the issuing of permits (IPPC Directive) and plans to improve air quality (Air Quality Framework Directive). In this, the two Directives are typical framework directives aiming to integrate substantively related, but separate pieces of EU environmental legislation.

Reflecting the integration regime's emphasis on negotiations and the inclusion of stakeholders, the IPPC and Air Quality Framework Directives feature various provisions concerned with informing the public. The IPPC Directive also provides for considerable stakeholder participation in the permit procedure and, in particular, in the so-called Seville Process for determining Best Available Techniques (BAT).[62] Further illustrating the negotiation-based decision-making mode associated with the integration regime, the IPPC Directive's definition of BAT, which was highly contentious at the time of the adoption of the Directive, was more an outcome of intensive negotiations involving Commission and national officials and experts as well as the European Parliament's Environment Committee than a reflection of voting rules in the Council.[63]

D. The Sustainable Development Regime

Although sustainable development is commonly understood to comprise an environmental, an economic, and a social dimension, the EU sustainable development regime is most directly concerned with environmental issues. But the concept of sustainable development as developed in the UN's 1987 Brundtland Report strongly emphasizes intergenerational justice and the time dimension. With its focus on *persistent* environmental problems, the EU

---

[59] Directive 90/313/EEC.   [60] Regulation 1210/90/EEC.
[61] Council Decision 97/872/EC.
[62] Koutalakis, C., *Regulatory Effects of Participatory Environmental Networks? The Case of the 'Seville Process'*, Paper prepared for the EUSA 9th Biennial International Conference, Austin, Texas, March 31–April 2, 2005.
[63] Cf. Héritier n. 43 and Héritier n. 45, pp. 25–7.

sustainable development regime reflects this emphasis. Persistent environmental problems are frequently linked to structural properties of the economic sectors causing the problems and are characterized by complex causal chains and delayed effects resulting in low visibility and uncertainty. They often also have a global dimension requiring internationally coordinated responses.[64] Climate change and the loss of biodiversity exemplify this type of problem.

The focus on long-term environmental problems at least partly explains the fact that many measures associated with the sustainable development regime are of a more strategic nature and do not amount to legally enforceable legislation. Often, these measures are adopted by the Commission and are then in one way or another supported by the Council. This applies, for example, to the 1998 Cardiff Process of environmental policy integration, the original and revised EU Sustainable Development Strategies of 2001 and 2006, and the Thematic Strategies which the Commission presented as a follow-up to the 2002 6th EAP. Although these measures do not need to be formally based on the Treaty due to their non-legislative character, Article 6 TEC provides a suitable rationale. That Article reflects the EU's commitment to sustainable development in that it requires environmental policy integration—the integration of environmental concerns into all Community policies. Environmental policy integration is, in turn, generally regarded as key to putting sustainable development into practice.[65] Although the sustainable development regime is more strongly associated with non-legislative measures than any of the other EU environmental governance regimes, legislation remains important. Legislative measures associated with the regime are usually based on Article 175 TEC.

Efforts to act on Article 6 TEC, in particular the Cardiff Process of environmental policy integration mentioned above, mark the emergence of the sustainable development regime. More specifically, the June 1998 European Council in Cardiff called on three sectoral Council formations—Agriculture, Energy, and Transport—to pioneer the development of environmental integration strategies. Among other things, this initiative reflected the commitments made at the global 1997 Rio+5 Summit which had reviewed progress in implementing the sustainable development agenda adopted at the 1992 Earth Summit. Additional political momentum stemmed from the 1995 EU accession of environmentally progressive countries, in particular Austria and Sweden, as well as the then recent change from a Conservative to an environmentally more ambitious Labour government in the UK—the country which held the Council presidency in the first half of 1998. Although the list of Council formations

---

[64] Cf. Jänicke/Jörgens, n. 1, pp. 169–71.
[65] Lafferty, W. and Jørgen, K., 'The Issue of "Balance" and Trade-offs in Environmental Policy Integration: How Will We Know Environmental Policy Integration When We See It?', EPIGOV Paper No. 11 (Berlin: Ecologic—Institute for International and European Environmental Policy, 2007), pp. 14–16, published on the Internet at: <http://www.ecologic.de/projekte/epigov/download-area.htm>; Lenschow, n. 14, pp. 6–7.

which developed environmental policy integration strategies according to the requirements of the Cardiff Process was subsequently significantly extended, the process came to a gradual standstill after the initial drafting of environmental integration strategies had been completed. Other initiatives, such as the EU Sustainable Development Strategy and the Thematic Strategies foreseen by the 6th EAP took centre stage.[66]

An important feature of the sustainable development regime is that when it comes to decision-making, implementation rather than the formulation of the original EU measures is critical. This is because the original measures tend to be highly underdetermined as a result of at least four factors. First, measures associated with the sustainable development regime frequently leave the formulation of concrete targets, limit values etc. to the implementation stage. Second, full implementation of measures often spans long periods of twenty and more years. Third, Member States and other bodies involved in implementation usually have wide leeway in deciding which particular approaches and instruments they want to use to implement EU requirements. Finally, even legislative measures associated with the sustainable development regime often do not impose legally binding obligations in all important respects.

Against this background, different types of learning in various technical and more political forums of scientific and administrative experts and stakeholders tend to constitute the most important decision-making mode of the sustainable development regime. The 2000 Water Framework Directive[67] provides an example. Implementation of that Directive is supported by the Common Implementation Strategy (CIS) which gives non-legally binding guidance to national and other lower-level authorities on implementation. The CIS consists of nested forums of state and non-state experts and stakeholders, ranging from the more political to the more technical and led by the regular meetings of the water directors, who usually head the water division in national environment ministries. Based on learning mechanisms such as the formulation and diffusion of guidance documents derived from national experience, testing of the documents, and mutual information and comparison, the CIS has been instrumental in implementing the vague substantive objectives as well as the more concrete procedural requirements of the Water Framework Directive—a process that is scheduled to be completed around 2020.[68]

Important political incentives for learning and subsequent adaptation stem from what has been called 'destabilisation' and a 'destabilisation regime'.[69] More

---

[66] Von Homeyer, n. 14; Pallemaerts et al., n. 10.
[67] Directive 2000/60/EC.
[68] Homeyer forthcoming, n. 14; Scott, J. and Holder, J., 'Law and 'New' Environmental Governance in the European Union', in de Búrca, G. and Scott, J., (eds.), *New Governance and Constitutionalism in Europe and the US* (Oxford: Hart Publishing, 2006).
[69] Sabel, C. F. and Zeitlin, J., 'Learning from Difference: The New Architecture of Experimentalist Governance in the EU', (2008) 14 *European Law Journal* 271–327, at 277 and 306.

specifically, the various actors participating in forums for the implementation of measures associated with the sustainable development regime may often have both significant incentives and opportunities to engage in information exchange and comparison and to search for, and commit to, solutions that are acceptable to their 'peers' because likely alternative outcomes, in particular inaction or top-down intervention by the EU or individual Member State governments, would amount to a major failure.

Again, the Water Framework Directive offers an example of this. As mentioned above, the CIS provides for multiple information exchange and mutual learning opportunities. In addition, several factors suggest that the main actors engaged in implementation of the Directive generally have a strong interest in avoiding implementation failure. First, the considerable body of pre-existing EU water regulation was widely perceived to be inadequate to address issues such as pollution from diffuse sources (in particular agriculture), implementation failures (for example of the Nitrates Directive),[70] and questions of water consumption, pricing, and management. Those latter issues, in particular, became increasingly pressing in the 1990s as a result of privatization in the water sector, which was partly a response to the high costs imposed by EU clean-up measures, such as the Urban Waste Water Directive, which required heavy investment in the 1990s and beyond.[71]

Second, were the WFD to fail it would be difficult to develop, and agree on, an alternative approach to water regulation in Europe. Detailed, top-down regulation at either EU or Member State level would have to overcome the cognitive constraints associated with top-down management of persistent environmental problems and related issues, such as diffuse sources of water pollution and management of water resources. Also, sovereignty concerns and the political sensitivity of the water supply severely limit the possibility of more direct EU intervention.[72] Similarly, an even more decentralized approach (or no alternative EU measures at all) would, among other things, face difficulties in addressing the critical interactions with the centralized EU Common Agricultural Policy as well as the growing economic dimension and multi-level character of water policy.[73]

Finally, the CIS has allowed national water directors to significantly expand their role in EU water policy. First, the CIS has equipped them with an additional—in some cases perhaps also alternative—transnational structure on top of their national administrations and constituencies providing expertise and stakeholder feedback. Among other things, this offers the water directors new opportunities to influence Commission initiatives at an early stage. Second, their

---

[70] Directive 91/676/EEC.
[71] Cf. Page, B. and Kaika, M., 'The EU Water Framework Directive: Part 2. Policy Innovation and the Shifting Choreography of Governance', *European Environment* 13 (2003) 328–43; Kaika, M., 'The Water Framework Directive: a new directive for a changing social, political and economic European framework', *European Planning Studies* 11 (2003) 303–20.
[72] Page and Kaika, n. 71, p. 13.
[73] Kaika, n. 71.

central role in the implementation of the WFD has allowed the water directors to also expand their influence beyond the EU to countries and international institutions which are interested in adopting measures which are very similar to the WFD or parts thereof.[74] If implementation of the WFD failed, the water directors would risk losing their newly gained EU and international roles. It should be noted that the fact that the water directors established the CIS very quickly after the WFD had come into force also suggests that they have a significant interest in implementing the WFD.[75]

Justifications of measures associated with the sustainable development regime often include a strong normative element referring to the principle of sustainable development. The fact that the EU adopted a host of environmental and environment-related strategies in the first years of the new millennium, in particular the EU Sustainable Development Strategy and the seven Thematic Strategies, seems to reflect attempts to create an EU-level normative framework which can be used to justify measures associated with the sustainable development regime.[76] Given this regime's focus on persistent environmental problems it is not surprising that such a framework includes strong normative elements. Because persistent environmental problems are associated with complex causes and delayed effects, uncertainty, and low visibility, the political utility of alternative justifications using scientific, technological, or pragmatic arguments is relatively limited if compared to the other EU environmental governance regimes.

The objectives of measures associated with the sustainable development regime tend to be similarly vague to those underlying the integration regime. However, they are often more long-term, ideally including target dates, and have a stronger normative component. In this, they reflect the regime's focus on persistent environmental problems as well as its stronger emphasis on normative justifications. For example, the Water Framework Directive's Article 4 requires Member States to achieve 'good water status' within a period of fifteen years. The Directive combines the normative concept of 'good status' with a target date and a long timeframe for implementation.

In terms of instruments, the sustainable development regime frequently relies on framework directives. However, these directives differ from the 'integrating' framework directives typically associated with the integration regime in that they are more 'reflexive' and incorporate the formulation of limit values and

[74] Cf. von Homeyer, I., 'Experimentalist Environmental Governance in the EU: Complex Challenges, Recursive Policy-making', International Implications, paper prepared for the 2008 Berlin Conference on the Human Dimensions of Global Environmental Change, 22–23 February 2008, Berlin, Germany.

[75] Cf. Pallemaerts et al., n. 10, p. 27.

[76] Baker, S., 'Sustainable development as symbolic commitment: Declaratory politics and the seductive appeal of ecological modernisation in the European Union', *Environmental Politics*, 16 (2007) 311–14, argues that the EU's commitment to sustainable development helps both to shape a European identity and to justify EU environmental policy and the European integration process more generally.

substantive measures to a considerable extent into the implementation process. This often happens even if the substantive measures are subsequently formally adopted as implementing decisions or as legislation, such as daughter directives. Put differently, while the specification of vague objectives largely relies on the EU legislative process or formal comitology in the case of the integration regime, this function is frequently taken over by processes established under the framework directives themselves in the case of the sustainable development regime.[77]

Various institutional and instrumental features help to enable the reflexive specification of vague objectives in these extra-legislative/extra-comitology processes. The development of common indicators and protocols is one of them. These instruments are reflexive in that they support meaningful cross-national comparison and mutual learning, and they are themselves subject to regular review and revision. This happens in forums composed of state and non-state actors and ranging from the more technical to the more political and which enable recursive, legally non-binding, yet authoritative decision-making.

Again, the Water Framework Directive (WFD) provides an illustration. For example, to support comparisons and ensure that 'good water status' means the same in all Member States despite a wide variation of conditions across the EU, the so-called 'intercalibration' exercise has been instituted. Intercalibration is expected to result in the designation of reference sites and the harmonization with respect to the comparability of results of ecological quality status assessment systems for surface waters.[78] This exercise is part of the WFD's Common Implementation Strategy (CIS). The CIS constitutes the extra-legislative/extra-comitology process which prepares, and often effectively takes, the critical decisions fleshing out the vague substantive objectives of the WFD.[79]

As mentioned above, the Cardiff Process of environmental policy integration, the EU Sustainable Development Strategy, and some of the Thematic Strategies and associated legislation, such as the recently adopted Marine Strategy Directive, provide additional examples of instruments associated with the sustainable development regime. Although all of these measures adopt a long-term, strategic perspective, some are more normative than others. This is particularly true for the EU Sustainable Development Strategy (EU SDS) which aims to 'identify and develop actions to enable the EU to achieve continuous improvement of the quality of life both for current and for future generations [...] ensuring prosperity, environmental protection, and social cohesion'.[80]

The EU SDS identifies several key challenges, such as climate change, sustainable transport, social inclusion, demography, and migration, and contains

---

[77] Cf. Pallemaerts et al., n. 10, p. 43–4.
[78] European Commission, Environmental objectives and intercalibration, published on the Internet at: <http://ec.europa.eu/environment/water/water-framework/objectives/index_en.htm>.
[79] Cf. Homeyer forthcoming, n. 14; Scott and Holder, n. 68.
[80] Council of the European Communities, 'Review of the EU Sustainable Development Strategy (EU SDS)—Renewed Strategy', 10117/06, Brussels, 9 June 2006, para. 5.

broad operational objectives and targets, as well as actions, addressing the key challenges. In addition to these operational objectives and targets, the type of instruments used also exemplifies the sustainable development regime. In particular, implementation of the EU SDS is supported by indicator development and use as well as specialized forums—the Commission's committee of national sustainable development coordinators and the independent European Sustainable Development Network (ESDN)—which support comparison and learning. However, these bodies appear to be significantly less resourceful and effective than the WFD's CIS and their influence on the substantive specification of the EU SDS tends to be quite small. In fact, the EU SDS's operational objectives and targets are not, by and large, original, but were adopted in various other political contexts which were, if anything, only weakly linked to the SDS. Specification of implementation measures similarly happens through processes other than those associated with the EU SDS.[81] Given the extremely broad scope of the EU SDS this is hardly surprising. As argued above, the justification of measures associated with the sustainable development regime which, in turn, generate substantive responses to address persistent environmental problems, appears to be the Strategy's primary function.

4. Conclusion

The analysis of the four EU environmental governance regimes and their emergence over time suggests that change has been endemic. However, as argued in the beginning, the older EU environmental governance regimes have only partly been replaced or modified. Instead newer ones have been added in what resembles a process of institutional layering. As a result, EU environmental governance is characterized not only by change, but also by considerable continuity. An estimate of the exact extent and of the causes of continuity and change would require a more detailed analysis than is possible in this chapter. However, notwithstanding its limitations, the analysis provides at least three potentially useful clues as to some of the factors which may have limited the extent of change. First, path-dependence might partly explain the persistence of certain EU governance patterns and corresponding environmental measures. This might be the case, for example, with the last remaining environmental issues, listed in Article 175 TEC, which still require unanimous decisions in the Council and with certain pieces of EU environmental legislation such as the Bathing Water Directive. The voting rules requiring unanimity have survived repeated efforts to change them, including recent intergovernmental negotiations on the Lisbon Treaty. While a revised Bathing Water Directive came into force in 2006, the Directive is little more than an update of its predecessor and was only adopted after earlier efforts in the 1990s to replace the

---

[81] Cf. Homeyer, forthcoming, n. 14.

original Directive had failed. Both the voting rules and the approach of the original Bathing Water Directive appear to be quite firmly entrenched.

Interaction of different governance regimes and associated measures might constitute a second cause of continuity which appears to be more closely related to institutional layering than to simple path-dependence. The integration regime seems to provide particularly instructive examples of this. More specifically, the regime and measures associated with the regime frequently 'integrate' pre-existing measures associated with other regimes, thereby enhancing the efficiency and effectiveness of these measures. This increases not only the utility and political viability of the pre-existing measures, but it also creates a certain degree of mutual dependence between the new integrative framework legislation and the pre-existing measures that are being integrated. Both effects tend to entrench the pre-existing measures.

Third, the differing problem focuses of the four EU environmental governance regimes suggest that the functional overlap among them is only partial. This provides a potential rationale for the coexistence of different governance regimes. Given partially differing problem focuses, even a governance regime which fits the more general political and environmental priorities particularly well in a given period is unlikely to prevail with respect to every single issue because the problem focuses of other regimes are likely to better reflect occasionally overriding issue-specific conditions and circumstances. For example, in a period in which environmental policy-making mainly focuses on persistent environmental problems and generally operates in ways associated with the sustainable development regime, acute problems might arise and be addressed in a technocratic manner and by top-down regulation using quality objectives associated with the environment regime. While the sustainable development regime would generally prevail in this scenario, there would also be some room for occasional measures associated with other regimes.

However, as regards continuing coexistence of EU environmental governance regimes, the sustainable development regime appears to differ in some respects from previous regimes. Although it seems unlikely that the sustainable development regime will end the coexistence of different regimes, certain developments as well as some of the characteristics of the regime suggest that it could be more likely to do so than previous regimes. The steep drop in the number of pieces of EU environmental legislation adopted since 2001 (see Figure 1.1) might indicate that the EU environmental governance regimes which strongly rely on legislation—the environment, internal market, and integration regimes—have been weakened.[82] Certain characteristics of the sustainable development regime might have contributed to this. In particular, measures associated with the

---

[82] The number of EU legislative acts adopted peaked in 2001—three years after the emergence of the sustainable development regime. However, the EU legislative process is very time-consuming so that the number of legislative proposals issued by the Commission must have peaked much closer to the time of the emergence of the sustainable development regime.

sustainable development regime usually define their own long-term environmental targets for comprehensive areas, and rely on an informal, learning-based decision-making mode which is capable of 'subverting' measures associated with other regimes. The WFD provides an example. First, it gradually repeals many pre-existing measures, such as the Drinking Water Abstraction Directive, and takes over their respective functions and tasks. Second, even if pre-existing measures are not repealed, the WFD tends to modify them in important respects so as to make them compatible with its comprehensive, long-term requirements. This is the case, for example, with the revised Groundwater Directive[83] which differs significantly from its 1980 predecessor because its requirements needed to be aligned with those of the WFD. Like the WFD some Thematic Strategies and associated framework directives can be associated with the sustainable development regime. However, these measures have only been adopted recently and the extent to which they will characterize EU environmental governance and override pre-existing regimes and measures in the future remains unclear.

Finally, the emergence of yet another EU environmental governance regime, which may already be under way, could be a factor constraining the proliferation of the sustainable development regime. In the past two years climate change has climbed to the very top of the EU political agenda. Although climate change has the characteristics of a persistent environmental problem, it is at the same time perceived as increasingly acute. Consequently, there are signs of a comeback of top-down legislative decision-making which, however, for the first time includes an important role by the European Council and the Commission President in EU environmental policy. Climate change therefore appears to be the first 'high politics' environmental issue. This opens up a considerable potential for mutual integration of climate change concerns with other policy areas, such as energy and security policy. Political justifications of measures echo this integration as they cover a broad range of arguments referring to environmental damage, health effects, economic implications, energy security, conflict scenarios etc. However, it seems clear that while such an emerging new EU environmental governance regime could weaken existing regimes, it could hardly replace them because of its exclusive focus on a particularly important, but nevertheless single, environmental issue.

---

[83] Directive 2006/118/EC.

2

Law and Governance of Water Protection Policy

Maria Lee

1. Introduction

The Water Framework Directive should mark a moment of real progress in the legal protection of water in the EU.[1] In keeping with increasing recognition of the complexity and interrelatedness of environmental challenges, the Directive takes an approach that is more environmentally integrated than its predecessors, with a greater awareness of ecological realities. With tough objectives and a series of clear deadlines, it makes significant demands of Member States.

Beside its daunting complexity, the Water Framework Directive is most striking for its ambitious and holistic environmental objectives. The ambition is closely connected to the seriousness with which the Water Framework Directive takes ideas of 'new' and 'multi-level' governance. In its approach to the governance of a complex environmental issue, the Directive expands on formal hierarchical relationships between different levels of government, and has the potential to disrupt regulatory assumptions at all levels. This chapter will begin by outlining the main features of the Directive,[2] before examining two different manifestations of 'new governance' embedded in the Directive: those governance techniques called on in the Directive itself, and those called on in the negotiation of implementation beyond the letter of the legislation. Both within and beyond the wording of the legislation, the Directive's efforts to influence and persuade rather than command and control are heavily reliant on the power of information and learning. The Directive demands the generation of diverse information to form the context for decision-making, and a range of actors contribute to that

---

[1] Directive 2000/60/EC establishing a framework for the Community action in the field of water policy [2000] OJ L327/1.
[2] For more detail on water legislation, see Ludwig Kramer, *EC Environmental Law* (Sweet and Maxwell, 2006). On new governance in the Water Framework Directive, see especially Scott, J. and Holder, J., 'Law and New Environmental Governance in the European Union' in de Búrca, G. and Scott, J., *Law and New Governance in the EU and the US* (Hart, 2006).

process. The terms of the Directive, conspicuously open-ended, are to be gradually filled out, both within the Member States, but also, in a novel process on which the Directive is silent, in collaborative networks of public and private actors from different levels of governance in the EU. This 'Common Implementation Strategy'[3] provides for a joint approach to implementation and has real potential for administrative and regulatory innovation. The Common Implementation Strategy, with its rhetoric of transparency and participation, is designed to enhance the accountability of Member States, not to their citizens generally, but to each other, to the EU itself and possibly to certain relatively narrow, expert 'publics'. Whilst this is welcome, it is important also to interrogate the relationship of the Common Implementation Strategy with the more localized publics affected by the implementation of the Water Framework Directive.

It must be unlikely that recourse to new governance is always deliberate and reflective. Vague and permissive language is sometimes the cost of agreement, and compromise can be expected to show on the face of legislation. It is also quite normal to see difficult decisions postponed for resolution outside the legislative process (through Commission 'interpretative' documents, negotiation between Commission and Member State, and comitology or future legislative processes). And certainly, water protection policy is no stranger to this. But although we should be cautious of the amount of work that ideas of 'tradition' might be doing in my discussion of new governance of water protection, there is something more interesting going on here than simple incompetence or compromise in the drafting of a directive. The Water Framework Directive is a self-conscious move to distinctive governance approaches, demanded by the extent of its substantive ambitions. For example, the nature of the regulatory units chosen (river basin districts, ecological quality) more or less mandate flexibility: an ecological assessment of water quality does not allow a 'one size fits all' approach to regulation, and the river basin district is inherently disruptive of ideas of EU, local, *and* national regulation. The flexibility in turn demands constraints.[4] New governance of water is more than just a poor relation to 'hard law'. For all its flaws and challenges, new governance allows the pursuit of objectives that simply could not have found a home in legislation. This is familiar from, for example, the use of the 'open method of coordination' in respect of aspects of social policy. But in the case of water, the formal legislative competence of the EU is not in doubt,[5]

---

[3] See the discussion in Scott and Holder, ibid.; Trubek, D. M. and Trubek, L. G., 'New Governance and Legal Regulation: Complementarity, Rivalry or Transformation' (2007) 17 *Columbia Journal of European Law* 539.

[4] See Joanne Scott's discussion of constraints on flexibility in Directive 1996/61 on integrated pollution prevention and control [1996] OJ L257/26 (IPPC Directive), 'Flexibility, "Proceduralization", and Environmental Governance in the EU' in Scott, J. and de Búrca. G. (eds.), *Constitutional Change in the European Union* (Hart Publishing, 2000).

[5] The role of the open method of coordination is of course more complex than simply a substitute for competence. There is now a large literature on the open method of coordination, but see for example the contributions to de Búrca and Scott, supra n. 2.

and new ambition in a long-standing and well established area of EU competence demands new approaches.

2. The Water Framework Directive

As in many areas of environmental protection, the long history of the regulation of water is characterized by fragmentation.[6] Types of water use, types of pollutant, types of potentially polluting activity, have been addressed one by one, possibly by different regulators with different approaches and different enforcement powers. This creates well recognised problems of regulatory coherence, as overlap and gaps emerge, but more fundamentally, also fails to appreciate the integrated nature of environmental problems. On top of this, water protection legislation in the EU has been generally poorly implemented and unresponsive to changing knowledge or environmental conditions.[7]

The Water Framework Directive takes a significant step towards the integration of water protection, addressing all types of water and all types of impact in a single regulatory framework, rather than addressing for example bathing water and shellfish waters, agricultural run off and urban waste water treatment, independently. As well as integrating the regulatory approach to water, the Water Framework Directive rationalizes the legislation, replacing a number of directives. The Directive has the potential to address diffuse as well as point-source pollution and the quantity as well as the quality of water. It also addresses hard engineering (for example canalization and diversion), and refers explicitly to links with land use.[8] One of the central features of the Directive is the regulatory unit of the 'river basin district', an ecological rather than administrative unit, framing regulation around hydrological complexities rather than administrative convenience. The Directive's use of 'ecology' is also a holistic approach to the water environment. The achievement of 'good ecological status' for surface waters is, as discussed below, one of the Directive's key aims. In this, the Directive looks beyond immediate human interests in the protection of water, considering living things dependant on water autonomously of human interest. This could be an early recognition of the value of ecosystems beyond their practical or aesthetic use to humans.[9] And finally, the Directive takes a 'combined approach' to standards, sidestepping the old debate between quality objectives or emission limits. The

---

[6] Although see Ben Pontin's interesting analysis of the move *away* from a starting point of integration during the late industrial revolution, 'Integrated Pollution Control in Victorian Britain: Rethinking Progress within the History of Environmental Law' (2007) 19 *Journal of Environmental Law* 155.

[7] See Jordan, A., 'European Community Water Policy Standards: Locked in or Watered Down?', *Journal of Common Market Studies* 37 (1999) 13.

[8] Particularly Annexes II and VII.

[9] See the discussion in Howarth, W., 'The Progression Towards Ecological Quality Standards' (2006) 18 *Journal of Environmental Law* 3.

'combined approach' means that emissions standards set in specified EU legislation shall be applied first, but if they are not adequate to meet environmental quality objectives, more stringent emissions standards shall apply.[10]

The Water Framework Directive contributes to the integration of the regulation of one environmental medium (water) and its relationship with land. It does not seriously attempt to integrate environmental considerations into other sectors of decision-making, as would be required by the 'integration principle' in Article 6 of the EC Treaty.[11] Even in this respect, however, it is hoped that the Directive will provide 'a basis for a continued dialogue and for the development of strategies towards a further integration of policy areas', in particular 'integration of protection and sustainable management of water into other Community policy areas such as energy, transport, agriculture, fisheries, regional policy, and tourism'.[12]

Article 4 contains the Directive's 'environmental objectives'. 'Good water status' is at the heart of the Directive, and the flexibility of this notion is at the heart of reliance on varied governance techniques. Article 4 requires all Member States to 'protect, enhance and restore' all bodies of water, with the *'aim'* of achieving 'good water status' by 2015, in respect of both surface water and ground water. Status embraces all impacts and so could be implicated, for example, by construction of any type, by reservoirs or canalization, as well as by introduction of pollutants. Member States *'shall* achieve compliance' with the Article 4 environmental objectives in protected areas.[13] And the general obligation to 'aim' to achieve good water status is backed up by a 'no-deterioration' principle, an obligation on Member States to prevent the deterioration of status of surface and ground water.[14] This is a potentially hard hitting obligation, but as discussed below, is subject to extensive exceptions.

The meaning of 'good water status' is both complex and unclear, and to a large degree left to the future, in a continuum of decisions at different levels of government. Surface water is judged by the criteria of 'chemical status' and 'ecological status'; ground water by reference to 'quantitative status' and 'chemical status'. 'Good ecological status' is to be determined in accordance with criteria set out in the lengthy Annex V. This is discussed further in the context of the Common Implementation Strategy, below, but without going into too much detail for now, ecological status addresses 'biological elements', 'hydromorphological elements supporting the biological elements' and 'physico-chemical elements supporting

---

[10] Article 10.
[11] Article 11 of the Treaty on the Functioning of the European Union, following the 2007 Lisbon Treaty.
[12] Recital 16.
[13] Article 4(1)(c). As well as areas protected under EU nature conservation legislation, 'protected areas' include areas designated for the abstraction of drinking water, areas designated for the protection of economically significant aquatic species, areas designated as recreational (including bathing) waters and areas designated as nitrate vulnerable zones, Annex IV.
[14] Article 4(1)(a)(i) and (b)(i).

the biological elements'.[15] Each type of water (rivers, lakes, coastal waters, etc, themselves further subdivided)[16] is assessed against a range of characteristics within these broad categories, for example composition and abundance of aquatic flora (a biological element), quantity and dynamics of water flow, structure and substrate of the river bed (hydromorphological elements), thermal conditions and nutrient conditions (chemical and physico-chemical elements). Whilst no acceptable thresholds or ranges are specified for these characteristics, ecological status is assessed basically by reference to 'naturalness'.[17] A body of water can be classified as 'high' ecological status if 'there are no, or only very minor, anthropogenic alterations' in these elements (biological, hydromorphological, chemical, and physico-chemical), compared with what is 'normally associated with that type under undisturbed conditions'. In good status waters, there are 'low levels of distortion resulting from human activity', deviating 'only slightly from those normally associated with the surface water body type under undisturbed conditions'.[18] Although the Directive aims for 'good' rather than 'high' status, the non-deterioration principle means that in principle measures to maintain 'unspoiled' waters will need to be taken. It is probably stating the obvious to point out that the notions of (at least) 'minor', 'normally', 'undisturbed', and 'only slightly' are unclear and subject to judgment.

Whilst a baseline of no interference is at least easy to identify (if not to achieve) in respect of contamination by artificial chemical substances, it is far more difficult to identify in respect of these other ecological characteristics, such as abundance of flora. The 'natural' water system is rare in Europe, where waters have been abstracted, channelled, diverted, deepened, and straightened for agriculture, urbanization, and industry, not to mention polluted by human, agricultural, and industrial waste, for very many centuries. Even the idea of a single pre-disturbance, pristine environment is a difficult one: William Howarth reminds us that 'the idea that nature, free from human interference, adheres to [an] unshifting equilibrium' is a myth.[19] Ecosystems are unstable and constantly changing. And even if something approaching pristine conditions can be identified, efforts to guarantee a particular abundance or diversity of species or flow of

---

[15] In the first instance, ecological status is the lower of biological, physico-chemical, and hydromorphological results. But independent assessment of physico-chemical and hydromorphological status is not necessary for every boundary—in some cases, the fact that the biological elements are consistent with a particular status (good, moderate, etc.) means that by definition so are the elements supporting them. For detail, see Common Implementation Strategy, Guidance Document No 13 *Overall Approach to the Classification of Ecological Status and Ecological Potential* (2005).

[16] So for example, in England and Wales, twelve lake types have been identified, Summary Report of the Characterisation, Impacts and Economic Analyses Required by Article 5: Thames River Basin District (DEFRA, 2005) p. 11.

[17] See the discussion in Howarth, supra n. 9.

[18] All these quotations are found in Annex V. Definitions are also provided of moderate, bad, and poor status.

[19] Howarth, supra n. 9.

water in the future are hugely problematic. The reasons for failure to meet such a standard are often at best only poorly understood, and will not always be due to human intervention.

'Good quantitative status', which applies to groundwater, is also elaborated in Annex V. The available groundwater resource must not be exceeded by the 'long-term annual average rate of abstraction'. Quantitative status is closely tied to the other environmental objectives of the Directive, and is also defined by reference to the ability of associated surface waters to achieve good status, and the significant diminution of the ecological and chemical quality of the associated surface waters.

Unlike ecological and quantitative status, chemical status can be defined basically by reference to specific numerical environmental quality standards.[20] For surface water, these standards are set out in other Directives listed in the Annex IX to the Directive, in 'daughter directives' on priority substances under Article 16, or 'under other relevant Community legislation setting environmental quality standards at Community level'. An obligation to address priority substances runs alongside the good chemical status obligation. Measures should be implemented with the 'aim' of 'progressively reducing' pollution from priority substances and 'ceasing or phasing out emissions, discharges and losses' of priority hazardous substances.[21] Good groundwater chemical status is, like quantitative status, defined in part by the ability of associated surface waters to achieve good status and avoid significant diminution of their ecological and chemical quality. Good groundwater chemical status is defined strictly, assuming that there are in principle no acceptable levels of contamination. The obligation in Article 4 is to take measures necessary both 'to prevent or limit the input of pollutants into groundwater' and 'to reverse any significant and sustained upward trend in the concentration of any pollutant resulting from the impact of human activity in order progressively to reduce pollution of groundwater'.[22] Direct

---

[20] Note that the comparison between imprecise ecological quality standards and precise public health standards observed under earlier Directives is likely to persist. So Howarth discusses the precise standards of nitrate pollution for human health purposes, as contrasted with the vaguer 'no significant deterioration' affecting flora or fauna or ecosystem, supra n. 9.

[21] Article 4(1)(a)(iv). Priority substances are identified under a process set out in Article 16(2) (highly controversial during the negotiation of the Directive). The list of priority substances can be found in Commission Decision 2455/2001/EC establishing the list of priority substances in the field of water policy and amending Directive 2000/60/EC OJ [2001] L331/1. Priority hazardous substances are identified in accordance with Article 16(3) and (6) for which measures have to be taken in accordance with Article 16(1) and (8). Hazardous substances are substances or groups of substances that are 'toxic, persistent and liable to bio-accumulate' and substances or groups of substances 'which give rise to an equivalent level of concern', Article 2(29) and (30). See European Commission, *Proposal for a Directive on Environmental Quality Standards in the Field of Water Policy* COM (2006) 397 final, and Commission's communication on the Council's Common Position (COM (2007) 871 final).

[22] Article 4(1)(b)(i) and (iii). See Directive 2006/118/EC on the protection of groundwater against pollution and deterioration OJ [2006] L372/19, a daughter directive foreseen by Article 17.

discharges of pollutants into groundwater are prohibited (subject to a range of exceptions).[23]

The notion of 'good water status' (good surface water and groundwater, good ecological, chemical, and quantitative status) is demanding as well as complex. 'Good water status' is subject to equally complex exceptions and extensions. The provisions reduce the ecological emphasis of the basic obligations through an explicitly anthropocentric rubric, providing considerable discretion for the balancing of environmental and other social, including economic, objectives.

The most far reaching exception to the good water status aim is in respect of an 'artificial or heavily modified body of water', which includes features like ponds, canals, reservoirs, and rivers deepened for navigation. Here, the obligation on the Member State is to 'protect and enhance' (rather than 'protect, enhance and restore'), with the aim of achieving (as well as good chemical status) 'good ecological *potential*'.[24] This is about reflecting 'as far as possible' what we would expect in the type of water body (river, lake, etc.) that the artificial or heavily modified body of water most closely resembles.[25] Member States can designate a body of surface water as artificial or heavily modified when the changes to the 'hydromorphological characteristics' of the water that would be necessary to achieve good ecological status would have 'significant adverse effects' on any of a broad range of interests, including not only 'the wider environment', but also human and economic purposes of navigation, recreation, drinking-water supply, power generation, irrigation, water regulation, flood protection, land drainage, and a catch-all 'other equally important sustainable human development activities'.[26] Artificial or heavily modified status applies only if the beneficial objectives of the water cannot 'reasonably' be achieved through a better environmental option for reasons of 'technical feasibility or disproportionate costs'.

It is important that the special status of heavily modified bodies of water is recognized. Restoring bodies of water engineered for, for example, power generation, flood defence, or port facilities to 'undisturbed conditions' would demand massive social and economic upheaval over a long period. And in any event, many ecologically valuable areas are highly dependant on human intervention, for example through drainage, flooding, or artificial water courses.[27] However, there is clearly the possibility for the basic standard of 'good ecological status' to be more of an exception than the rule in practice.[28] Whilst the very broad type of information to be considered in the designation of an artificial or

---

[23] Article 11(3)(j).
[24] Article 4(1)(a)(iii).
[25] Like good ecological status, detail can be found in Annex V.
[26] Article 4(3).
[27] See Howarth, supra n. 9.
[28] The percentage of heavily modified or artificial water bodies varies enormously between Member States, from under 2% (provisionally) in Ireland and Latvia, to 95% in the Netherlands, see Commission Working Document *Accompanying document to COM* (2007) 128 final SEC (2007) 363.

heavily modified body of water is set out (including especially technical feasibility and the proportionality of cost) quite how that information should be weighed and dealt with is left open.

There is a range of other exceptions in Article 4. Under Article 4(4), extensions to the 2015 deadline are possible if Member States conclude that the necessary improvements cannot 'reasonably' be achieved on time for reasons of 'technical feasibility', 'disproportionate expense', or because 'natural conditions do not allow timely improvement in the status of the body of water'.[29] This extension is potentially open-ended, because although in principle limited to two updates of the River Basin Management Plan (updated every six years), that does not apply 'in cases where the natural conditions are such that the objectives cannot be achieved within this period'.[30] Under Article 4(5), less stringent environmental objectives are acceptable in respect of bodies of water when either they are 'so affected by human activity' or 'their natural condition is such', that the achievement of good water status would be 'infeasible or disproportionately expensive'. This exception only applies if 'the environmental and socioeconomic needs served by such human activity cannot be achieved by other means, which are a significantly better environmental option not entailing disproportionate costs', and subject to the non-deterioration principle. Article 4(6) provides for 'temporary deterioration in the status of bodies of water'[31] in cases of force majeure that could not reasonably have been foreseen ('in particular extreme floods and prolonged droughts'),[32] and accidents which again could not reasonably have been foreseen. And, matching the provisos that apply to artificial or heavily modified bodies of water that exist at the time of implementation, Article 4(7) provides a more stretching derogation for new changes, specifically 'new modifications to the physical characteristics of a surface water body or alterations to the level of bodies of groundwater', either if the modification is for reasons of 'overriding public interest',[33] or if the benefits 'to the environment and to society' of achieving good water status or non-deterioration are 'outweighed by the benefits of the new modifications or alterations to human health, to the maintenance of human safety or to sustainable development'.[34] Conditions of 'technical feasibility' or 'disproportionate cost' again apply, and 'all practicable steps' must be taken to mitigate the adverse impact. All of these provisions are

---

[29] Article 4(4)(a)(i)–(iii).
[30] Article 4(4)(c).
[31] Article 4(6).
[32] Not only are extreme floods and droughts reasonably foreseeable, but they have in fact been foreseen in general terms by the Directive. So reasonable foreseeability arguably implies degrees of likelihood, and consideration of how reasonable it is to provide for unlikely events.
[33] This is undefined, but is familiar from Directive 1992/43 on the conservation of natural habitats and of wild fauna and flora [1992] OJ L206/7.
[34] This applies to either a failure to reach good status or breach of the non-deterioration principle. Deterioration of a body of surface water from high to good status (only) is permissible for 'new sustainable human development activities'.

subject to an obligation that their application does not 'permanently exclude or compromise' the Directive's objectives in respect of other bodies of water, that it is consistent with other environmental legislation, and also that protection is not reduced relative to earlier legislation.[35]

The potential breadth of these exceptions is evident, and whilst they are heavily hedged about with conditions and provisos, the openness of many of the key terms enhances national discretion in implementation. So for example, although the notion of disproportionate costs clearly builds on the economic analysis discussed below, it is nevertheless highly contingent in its application, with real discretion as to what factors should go into the calculation of cost and benefit, or the appropriate relationship between benefits and cost. The balance between the good status and non-deterioration obligations and the rest of Article 4 is far from clear cut. The UK Government refers to 'alternative objectives' (alternative to good status and non-deterioration), and requires the Environment Agency to 'make full use' of them: 'they are an integral part of the WFD objectives and their use should be a normal part of river basin planning'.[36]

Article 4 of the Directive is characterized by enormous flexibility, and that flexibility is central to the governance approach discussed in the rest of this chapter. Water legislation had been controversial for many years before the advent of the Water Framework Directive, and disagreement on the scientific or economic basis for a standard has often been met by ad hoc derogations and extensions, vague and permissive language, or the postponement of certain difficult decisions until after agreement of the legislation.[37] The Water Framework Directive was a controversial and hard fought piece of legislation. The stakes were high, and the tortuous path to agreement provided multiple, and novel, opportunities for non-governmental influence on the final Directive.[38] It is certainly arguable that some of its ultimate vagueness is down to difficult compromise. But the flexibility or vagueness of the Directive's environmental objectives goes hand in hand with a progressive and holistic overall approach to the water environment. This Directive attempts to do something that EU regulation has not done before, and consider the ecology of water in a sophisticated way, on the basis of new administrative units, and assuming information that simply does not yet exist. The flexibility of the Directive, and associated governance techniques within and beyond the terms of the legislation, is at least in part a considered reflection of the very different ecological conditions in different bodies of water

---

[35] Article 4(8) and (9).
[36] Department for Environment, Food and Rural Affairs, *River Basin Management Planning Guidance* (DEFRA, 2006), para. 9.9.
[37] See Jordan, supra n. 7; Kramer, supra n. 2.
[38] See Kaika, M. and Page, B., 'The EU Water Framework Directive: Part 1. European Policy Making and the Changing Topography of Lobbying', *European Environment* 13 (2003) 314. The Water Framework Directive was negotiated in the early days of co-decision, and was the first time the conciliation committee was used.

and different parts of the EU, and the impossibility of a single standard suitable for all purposes:

> There are diverse conditions and needs in the Community which require different specific solutions. This diversity should be taken into account in the planning and execution of measures to ensure protection and sustainable use of water in the framework of the river basin. Decisions should be taken as close as possible to the locations where water is affected or used. Priority should be given to action within the responsibility of Member States through the drawing up of programmes of measures adjusted to regional and local conditions.[39]

3. Flexibility, Decentralization, and Control

Rather than clear and obviously binding directions, the Water Framework Directive is dominated by tools to influence the mind of the decision-makers. It is perhaps paradoxical to say that the notion and application of 'good water status' is open-ended and uncertain, when there are pages of complex guidance on the subject in the Directive. But whilst there are some clear, numerical standards in certain areas (especially chemical status), much of the legislation consists of setting out the factors that must be taken into account, in fairly general terms, without explicitly acceptable and unacceptable boundaries. This applies at every stage from the assessment of the different parameters for good ecological status, to the identification and weighing of costs and benefits for the application of exceptions. But flexibility is subject to techniques of governance that constrain and attempt to influence decision-making. The range of tools that Member States must use in their decisions, and the range of considerations that they must take account of, could be cut in many different ways. For the purpose of analysis here, I turn first to the way in which the planning obligations of the Directive force the generation of information and knowledge, followed by the closely related issues of economic instruments and public participation. The next section considers the mechanisms of implementation that are not apparent on the face of the legislation.

1. Techniques of governance (i): information forcing

There is a distinct emphasis in the Water Framework Directive on mechanisms that encourage institutions to learn, to revisit decisions, and to generate and absorb new information. This is a striking aspect of the Common Implementation Strategy, discussed further below, but obligations in the Directive itself to generate or just to consider particular types of information and knowledge are equally significant. Most obviously, the Directive contains detailed lists of the sorts of considerations that must be taken into account in decision-making. There is little guidance on the level of importance to attribute to different factors, or

---

[39] Recital 13.

compulsion on acceptable and unacceptable ranges, but the framework within which decisions are made exerts an influence from the legislator. In Annex V, for example, as discussed, there is a wealth of criteria against which the status of water bodies is assessed, calling on disciplines from chemistry to palaeoecology. The monitoring obligations in the Directive are also onerous.[40]

The Water Framework Directive contains potentially influential planning and reporting obligations, most importantly in the form of the River Basin Management Plan.[41] The River Basin Management Plan obliges Member States to make public commitments and provide public explanations, allowing scrutiny of the approach taken to the river basin district. There are obligations of public participation in the development of the Plan, as discussed below, and the Plan is also sent to the Commission and other Member States, providing a form of peer review. The Planning process is an important balance to the flexibility of the Directive's key objectives, since it provides an incentive to heed the spirit as well as the letter of the legislation. So for example, 'good water status', it will be recalled, is simply an 'aim'.[42] But the River Basin Management Plan must contain a summary of the programme(s) of measures[43] that will be put in place to achieve that aim, and when a body of water does not meet or is at risk of failing to meet good water status, an explanation has to be provided, both of the failure and of the extra monitoring and remedial obligations that kick in when a body of water does not meet or is at risk of failing to meet good water status.[44] The application of exceptions also has to be explained, a striking procedural inhibition in respect

---

[40] Article 8.
[41] Article 13. The first is due by 2009, and they must be updated every six years. The Water Framework Directive also contains familiar obligations to report on performance to the Commission, which then reports on the information received. Article 18(5) provides for a conference of interested parties on the report. The first conference took place on 22–23 March 2007, with over 400 participants. The already complex planning task is in principle supplemented by the obligations in Directive 2001/42 on the Assessment of the Effects of Certain Plans and Programmes on the Environment [2001] OJ L197/30. For detailed analysis of the application of the Strategic Environmental Assessment Directive, see William Howarth, 'Substance and Procedure under the Strategic Environmental Assessment Directive and the Water Framework Directive' in Holder, J. and McGillivray, D. (eds), *Taking Stock of Environmental Assessment* (Routledge-Cavendish, 2007). The UK Government interprets the Directives to mean that an 'environmental report' on the Plan must be produced during its preparation, see DEFRA, supra n. 36.
[42] The Court of Justice is not likely to leave the application of these concepts to the unfettered discretion of the Member States. In Case C-32/05 *Commission v Luxembourg* [2006] ECR I-11323, the Court stated that 'Member States are to take the necessary measures to ensure that certain measures formulated in general and unquantifiable terms are attained, whilst leaving the Member States some discretion as to the nature of the measures to be taken', para. 43, and explicitly compared (para. 40) 'most of the provisions' of the Water Framework Directive to Case C-365/97 *Commission v Italy (San Rocco)* [1999] ECR I-7773. In that case, the Court states that whilst Article 4 of Directive 1975/442 on Waste [1975] OJ L194/39, amended by Directive 1991/156 [1991] OJ L078/32 'does not specify the actual content of the measures that must be taken … it is nonetheless true that it is binding on the Member States as to the objective to be achieved whilst leaving to the Member States a margin of discretion in assessing the need for such measures', para. 62.
[43] Article 11.
[44] Article 11(5).

of delay or non-application of the Article 4 'good status' and 'non-deterioration' obligations. Notwithstanding the aspirational nature of some of the legislative language, there are meaningful consequences to failure to comply.

Obligations to explain and justify decisions are potentially powerful tools of accountability,[45] politically as much as legally. A further feature of the Water Framework Directive's planning obligations is their organization around the ecological unit of the river basin. This reinforces the ecological integration of the Directive, and not only fits neatly with subsidiarity's emphasis on the ecologically most appropriate level of regulation, but should lead to an integrative form of policy learning, eventually forcing links to be made on the river basin scale.[46] The river basin district is of particular interest in an EU context, given that river basins cross national borders.[47] The actual legal demands on cross-border administration are, however, relatively weak. For an international river basin district entirely within the EU, 'Member States shall ensure coordination with the aim of producing a single international river basin management plan'. Lack of success is anticipated, with provision for Member States simply to 'produce river basin management plans covering at least those parts of the international river basin district falling within their territory'.[48]

The River Basin Management Plan is also a central tool in the Water Framework Directive's focus on learning about river basin districts, inherent in the most forward looking aspects of the Water Framework Directive. The River Basin Management Plan requires the generation and collection of information on every river basin district. The analyses of the river basin district required under Article 5 are vital: 'an analysis of its characteristics', 'a review of the impact of human activity on the status of surface waters and on groundwater', and 'an economic analysis of water use'.[49] These three categories of information underpin the implementation of the Directive. Information is rarely neutral, and assumptions made in the absence of knowledge can be made distinctly uncomfortable by the generation of information. The 'characterization' of the waters includes a requirement to identify early in the process water bodies at risk of failing the environmental quality objectives,[50] and this has an intuitively obvious potential impact on what action the regulator (and outsiders) will think necessary.

---

[45] See Sabel, C.F. and Simon, W.H., 'Epilogue: Accountability Without Sovreignty' in de Búrca and Scott, supra n. 2.

[46] Flynn, B. and Kröger, L., 'Can Policy Learning Really Improve Implementation? Evidence From Irish Responses to the Water Framework Directive', *European Environment* 13 (2003) 150.

[47] See also Macrory, R. and Turner, S., 'Participatory Rights, Transboundary Environmental Governance and EC Law' (2002) 39 *Common Market Law Review* 489.

[48] Article 13(2). Where the international river basin district extends beyond the EU, 'Member States shall endeavour to produce a single river basin management plan, and, where this is not possible, the plan shall at least cover the portion of the international river basin district lying within the territory of the Member State concerned', Article 13(3).

[49] Subject to detail in Annexes II and III.

[50] Annex II.

Identifying the main human pressures on those water bodies is similarly of immediate regulatory interest.

Less obviously, the 'economic analysis of water use'[51] for each river basin district, even in the absence of detail on what should be done with this information, sets up a particular framework for decision-making. The scope of the economic analysis is very much in the hands of the Member States: the amount of data, how it is collected, the scope of the benefits and costs taken into account, are all a matter of professional judgement in the carrying out of the economic analysis.[52] The generation and availability of information nevertheless has the potential to disrupt assumptions and create regulatory pressures. Economic information appears to be seen in part as a way gradually to enhance the likelihood of environmentally beneficial but possibly unpopular national action. In particular, the references in the Directive to the use of economic instruments, discussed below, suggest that economic analysis could be used to achieve the internalization of environmental costs of water use. Any indirect subsidies of water exploitation will at least be in the open. But on the other hand, economic analysis in the Directive seems to be equally concerned with identifying the least costly way of meeting particular objectives, with a subtext that regulators and regulatees must not be overburdened.[53] So Article 11 (which requires Member States to put together a 'programme of measures' for the implementation of the Directive) refers explicitly to the need to 'make judgements about the most cost-effective combination of measures in respect of water uses to be included in the programme',[54] and Article 4 contains frequent reference to disproportionate costs, as discussed above. Questions of cost-effectiveness and disproportionate cost are supremely open to individual national judgements in all but perhaps the most extreme cases. But the existence of economic information (in many Member States for the first time) means that water protection policy will be pursued in a context of awareness of the costs of failing to regulate, but also the economic costs of regulating. The Water Framework Directive could turn out to

---

[51] 'Water use means water services together with any other activity identified under Article 5 and Annex II having a significant impact on the status of water', Article 2(39). (See text at n. 60 infra on 'water services'.)

[52] Although the Common Implementation Strategy has produced Guidance Document No 1, *Economics and the Environment—The Implementation Challenge of the Water Framework Directive* (2003), which provides some general guidance, and emphasizes the centrality of economic analysis to implementation.

[53] The task of economic analysis itself is likely to be burdensome to some degree, and the 'costs associated with collection of the relevant data' are explicitly a factor to take into account in deciding how much data to use, Annex III.

[54] Annex III. Annex III requires the economic analysis to contain 'enough information in sufficient detail' to carry out certain tasks, including applying the principle of cost recovery. The economic analysis on which cost recovery rests is drawn up: 'taking account of long term forecasts of supply and demand for water in the river basin district and, where necessary:
- estimates of the volume, prices and costs associated with water services, and
- estimates of relevant investment including forecasts of such investments'.

be controversially innovative in encouraging the routine consideration of economic issues in environmental legislation.

Whilst the first River Basin District Management Plans are not due until 2009, summary reports on Article 5 had to be prepared by 2005. The ambition of the obligation to generate information at a local level can be seen by a simple glance at some of the reports.[55] The Thames report is characterized as a risk assessment and a first step rather than 'a measure of the true status of waters at the present time'; the results will be used to 'steer' further work. The shortage of data facing those compiling the Thames report is also clear: 'the impact of a number of pressures relevant to the analysis have not been routinely monitored across the UK or are not yet well understood, eg the impact of hydromorphology pressures on ecology'.[56] Moreover, the meaning of even such key terms as 'good status' is still being worked out.

Data gaps, poorly defined objectives, and sheer complexity all mean that this initial reporting under Article 5 fell short of the very ambitious appearance of the Directive. The Commission confirms that 'most Member States put considerable effort into this first analysis', but that quality and detail vary.[57] But this need not be seen as an unadulterated failing, since it has contributed to an expectation of provisionality and learning as implementation of the Water Framework Directive progresses and the knowledge base improves. The Water Framework Directive is in important respects information 'forcing', demanding the production of information that does not exist and is not easy to obtain. No one seems to have expected the full range of information to be available in time for the Article 5 reports, but the Directive's requirements are a way to stimulate local learning, which is then passed on to the EU and to regulators of other river basins.

The Water Framework Directive demands the generation of a degree of information and knowledge on river basin districts that is by no means easily available. This is far from a trivial matter, and its significance, especially in the context of innovative and ambitious objectives, should not be underestimated. This information is then presented to the public and shared with peers, providing both external pressure to proper compliance and help on good practice.

2. Techniques of governance (ii): economic instruments

Obligations to put in place systems of prohibitions and prior authorization have always been at the heart of EU environmental regulation, and although the Water Framework Directive is self-conscious in its use of innovative techniques of

---

[55] See for example the Thames Report, supra n. 16. The reports for England and Wales can be found at <http://www.defra.gov.uk/environment/water/wfd/article5/index.htm>.

[56] Ibid., p. 2.

[57] European Commission, *Towards Sustainable Water Management in the European Union—First stage in the implementation of the Water Framework Directive* 2000/60/EC COM (2007) 128 final, p. 7.

regulation, it is far from eschewing 'command' or hierarchy. Article 11 requires the Member States to put in place a 'programme of measures' for the implementation of the Directive. Among the 'basic measures' that 'shall' be included in this programme are prior authorization or prohibitions on the abstraction and impoundment of fresh surface water, of augmentation of water, point-source discharges 'liable to cause pollution', diffuse sources liable to cause pollution, and 'any other significant adverse impacts'. The Member States are thus required to exercise their own command capacity.

However, 'economic or fiscal instruments' of regulation are explicitly mentioned as a 'supplementary' measure of implementation, which 'shall' be included in the programme of measures 'where necessary'.[58] Article 9 goes a little further, providing that Member States shall 'take account' of the 'principle of recovery of the costs of water services'.[59] Water services are those services that provide 'for households, public institutions or any economic activity' both the provision of and discharge into water,[60] and explicitly in Article 9 include 'environmental and resource costs', 'in accordance in particular with the polluter pays principle'.

This part of Article 9(1) is an apparently soft-edged direction to the Member States on how to reach decisions in the implementation of the Water Framework Directive. They must take into account a certain category of information, although again there is no compulsion on what degree of importance to attribute to this category of information, and certainly no duty to apply 100% cost recovery. The duty to take into account is though disruptive of policy approaches that see water as 'free'. And the second sentence of Article 9(1) provides slightly more specific demands in respect of economic instruments: Member States are to 'ensure' by 2010 that 'water-pricing policies' will provide 'adequate incentives for users to use water resources efficiently', ensuring 'an adequate contribution of the different water uses', which must be 'disaggregated into at least industry, households and agriculture'. This requires the use of a specified regulatory tool. The identification of 'adequate' incentives and the degree of disaggregation are open-ended, and the detail of pricing could be hugely varied. As with Article 4, the Member States are directed but not without flexibility. But this is an undisguised effort to prohibit flat rate water charges and certain cases of water subsidy. Water pricing has a number of possible objectives, and the Directive settles for incentivizing the efficient use of water. The reference to 'efficient use' is a reminder that certain land uses are currently getting their water too cheaply and that allowing others to bear their external costs (water infrastructure, environmental costs) can make uneconomic activities appear profitable. But achieving the environmental objectives of the Directive may require more than efficient

---

[58] Article 11(4) and Annex VI, Part B (iii).
[59] For a detailed analysis of Article 9, see Bloch, H., 'European Water Policy and the Water Framework Directive' (2004) *Journal for European Environmental & Planning Law* 170.
[60] Article 2(38).

use.[61] The costs necessary for behaviour change are not necessarily the same as the external costs of water.

The negotiation of Article 9 was highly contentious. Water pricing has potentially disruptive socio-economic effects, most obviously a potentially regressive distributional effect between rich and poor households, but also between economic, industrial, and agricultural activities. Most notoriously, Spain resisted the entire Directive out of concern for its agricultural sector, heavy users of cheap water, and similar concerns applied around southern Europe.[62] Disagreement over the very concept of water pricing might explain the flexible language, and the discretion in Article 9 to have regard to 'social, environmental and economic effects of the recovery as well as the geographic and climatic conditions of the region or regions affected'. More specifically, the final paragraph of Article 9 thoroughly repatriates discretion on this topic. According to this final paragraph, Member States can decide 'in accordance with established practices' simply not to apply paragraph 1, second sentence (on water pricing and disaggregation). This option is only applicable to a 'particular water use activity', not across the board, and only 'where this does not compromise the purposes and the achievement of the objectives of this Directive'.[63] There is not a completely free rein for Member States, and as with the exceptions and extensions in Article 4, Member States have to report their reasons in the River Basin Management Plan. But the Directive's commitment to water pricing is certainly ambiguous.[64]

Whatever the ambiguity, the Water Framework Directive successfully gets water pricing onto the agenda and into the regulatory process. Decision-making in the Member State will be carried out in the context of information generated under Article 5 on the cost of water, exhortation under Article 9 on water pricing, and any direct or indirect subsidy of water intensive activities will be in the public domain. And the Commission intends to 'make the use of economic instruments a priority', through 'exchanges of information'.[65] There is likely to be increasing pressure towards the national use of economic instruments.[66]

---

[61] Massarotto, A., 'Water Pricing and Irrigation Water Demand: Economic Efficiency versus Environmental Sustainability', *European Environment* 13 (2003) 100 argues that pricing will make irrigation more efficient, in that it will be aimed at high value crops, but not necessarily more sustainable.

[62] The north/south division is not as clear cut as this suggests. See generally the discussion in Page, B. and Kaika, M., 'The EU Water Framework Directive: Part 2. Policy Innovation and the shifting Choreography of Governance', *European Environment* 13 (2003) 328.

[63] The state funding of pollution clean-up and water provision is still allowed—the funding of 'particular preventive or remedial measures' is permitted, Article 9(3).

[64] And note also Recital 9 of the Groundwater Directive, supra n. 22, which provides that: '[t]he protection of groundwater may in some areas require a change in farming or forestry practices, which could entail a loss of income. The Common Agricultural Policy provides for funding mechanisms to implement measures to comply with Community standards...'

[65] Commission, supra n. 57, p. 12.

[66] Note also that Directive 2004/35 on Environmental Liability with Regard to the Prevention and Remedying of Environmental Damage [2004] OJ L143/56 imposes an obligation on polluters to restore 'water damage', that is 'any damage that significantly adversely affects the ecological,

3. Techniques of governance (iii): public participation

One of the key governance techniques imposed on Member States through the Water Framework Directive is public participation. It is now perfectly routine for new environmental legislation to at least acknowledge the importance of public participation. The Directive was agreed in 2000, two years after the influential UNECE Aarhus Convention,[67] which provides certain rights of public involvement in environmental decision-making, was signed by all Member States and the EU. The turn of the century was also a time of high hopes for a future of 'participatory' or 'deliberative' democracy for the EU well beyond the environmental context, forming an element of the Commission's rethinking of 'European governance' more generally.[68]

Public participation potentially fills many roles. In the context of vague objectives such as those in the Water Framework Directive, public participation goes with the recognition that many important decisions will be made at the implementation stage. The normal democratic credentials found in legislative processes (as controversial as they might be at EU level) do not provide adequate legitimacy for this exercise of discretion. More instrumentally, participation in implementation can keep Member States on the straight and narrow. Broad involvement can also contribute to a complex and multi-faceted decision, for which all the necessary knowledge and information is unlikely to be found in a single institution. In this respect, public participation fits nicely with the determination of the Water Framework Directive to generate information and learning on water. Public participation can provide information and knowledge to decision-makers that simply cannot be found in a single bureaucracy, however expert. The sheer complexity of river basin management demands external involvement.[69]

Article 14 of the Directive provides for participation in implementation of the Directive. Member States are to 'encourage' the 'active involvement' of 'all interested parties'. The notion of 'active involvement' is potentially significant, moving beyond simple provision of information and an 'opportunity to object', and suggesting that participation should move up the 'ladder of participation' to

---

chemical and/or quantitative status and/or ecological potential ... of the waters concerned'. The Environmental Liability Directive has mixed objectives, but is partially premised on economic grounds—the Directive itself states that financial liability is imposed 'in order to induce operators to adopt measures and develop practices to minimise the risks of environmental damage so that their exposure to financial liabilities is reduced', Recital 2.

[67] Convention on Access to Information, Public Participation in Decision-making and Access to Justice in Environmental Matters (UNECE 1998).

[68] See European Commission, *White Paper on European Governance* COM (2001) 428 final, and associated documents.

[69] In the UK, there is an expectation that drawing up a Plan will involve a 'partnership with the range of public, private and voluntary sector organisations that will be affected by the [River Basin Management Plans]', DEFRA, supra n. 36, p. 8.

shape and influence decisions.[70] In the absence of information on how (or how much) to encourage, and quite how 'active' the involvement should be, much is left to the individual Member States, who may turn to tried and tested methods of consultation. The harder, clearly mandatory, language of Article 14 only explicitly demands information provision and an opportunity to comment in writing. The obligation to encourage active involvement moreover applies only to 'interested parties', although 'the public, including users' are entitled to be provided with a range of information, on which they are allowed at least six months to comment in writing in order to 'allow active involvement and consultation'. The language is far from clear,[71] but whilst there is no suggestion that public participation should be limited, the temptations of efficiency and ease of implementation create obvious incentives for regulators to close down the scope of involvement. By way of example, examining the environmental restoration of two English rivers in an urban environment, Sally Eden and Sylvia Tunstall illustrate the complexity of the socio-economic context of river ecosystems:

> the [River] Alt was perceived by local people not a as a green corridor of biodiversity but as a fast-getaway route for potential criminals from other socioeconomic groups 'outside' the estate, as well as a risk to small children playing in the area.[72]

An appropriate form of public involvement could engage with these sorts of concerns early in the process. But whether regulators will find the time and other resources required for meaningful engagement with the inhabitants of urban estates is at least debatable.

Although Article 14 applies to the Directive's provisions generally, the primary locus for public participation is around the River Basin Management Plan. Individuals are more likely to engage with the range of issues that are of concern to them the more concrete, the closer, and the more immediate proposals are. The river basin scale could to this extent be a barrier to broad participation. More positively, however, river basin management implies transboundary environmental regulation along ecologically relevant units. Whilst the Directive is silent on the subject, this should similarly imply transboundary participation, and potentially powerful ideas of 'European environmental citizenship'.[73] The Strategic Environmental Assessment Directive adds clearer obligations of transboundary participation where a plan is likely to have significant transboundary effects.[74]

---

[70] Arnstein, S.R., 'A Ladder of Citizen Participation', *Journal of American Planning Association* 36 (1969) 216.

[71] That second 'active involvement' could refer back to the 'interested parties'.

[72] Eden, S. and Tunstall, S., 'Ecological versus Social Restoration? How Urban River Restoration Challenges but also Fails to Challenge the Science-Policy Nexus in the United Kingdom', *Environment and Planning C: Government and Policy* 24 (2006) 661, 669.

[73] Macrory and Turner, supra n. 47. See also Christopher Hilson, 'Greening Citizenship: Boundaries of Membership and the Environment' (2001) 13 *Journal of Environmental Law* 335.

[74] Above n. 41, Article 7.

River Basin Management planners have obvious incentives to engage with external expertise on water. Whilst providing case studies to highlight the potential of innovative and intensive techniques, guidance on public participation focuses on 'stakeholders', and sees the role of participation as being to improve implementation in a fairly narrow sense.[75] And in the UK, for example, whilst broader participation is discussed, the emphasis is very much on seeking consensus in relatively small 'liaison' groups.[76] Support for the idea that technical expert discourse is alone sufficient for either legitimacy or effectiveness is however rare now in the EU. But putting this into practice is obviously enormously difficult and an ongoing process. The Water Framework Directive may even lead to two tiers of 'participation': 'active involvement' for a relatively narrow group of 'interested parties', which might include environmental and business interest groups; and a 'mere' opportunity to provide written comments for the more general public. The Commission acknowledges that the decisions in the River Basin Management Plan 'will involve balancing the interests of various groups', but participation is something of an afterthought: 'The economic analysis requirement is intended to provide a rational basis for this [balancing of interests], but it is essential that the process is open to the scrutiny of those who will be affected'.[77] That presentation does not bode well for the ability of non-expert value-based debate to compete with the economic analysis.

Finally, and more positively, the obligation in Annex VI to include in the River Basin Management Plan 'a summary of the public information and consultation measures taken, their results and the changes to the plan made as a consequence', is an important reason giving obligation. It not only concentrates the mind of the decision-maker but also allows public and peer scrutiny of participatory measures.

4. The Substantive Filling Out of the Water Framework Directive: Governance Beyond the Legislation

The Water Framework Directive is a complex and technically demanding piece of legislation and, as a framework directive, leaves a great deal of detail to be decided later. Possibilities for the filling out of the details of the legislation include traditional daughter directives[78] and administrative implementation by the Commission with national involvement through comitology.[79] Even in respect of

---

[75] The Common Implementation Strategy has drawn up extra-legislative guidance: Guidance Document No 8, *Public Participation in Relation to the Water Framework Directive* (2003).
[76] DEFRA, supra n. 36, paras 11.14–16.
[77] Introduction to the Water Framework Directive on the DG environment website, <http://ec.europa.eu/environment/water/water-framework/info/intro_en.htm>.
[78] For example the Groundwater Directive, supra n. 22.
[79] The Comitology (or Article 21) Committee is provided with competence to adapt 'to scientific and technical progress' a number of elements of the Directive, see especially Article 20.

comitology and legislation, bodies of governance have been put in place that are not apparent on the face of the legislation. The Commission has established 'multi-stakeholder Consultative Fora' for the drawing up of daughter directives and the further development of Community water policy, including the 'Expert Advisory Forum (EAF) on Priority Substances and Pollution Control', supposed to be the key working group in respect of priority substances. The intention seems to be that these bodies will be more open to external (environmental groups, industry, European Parliament) participation and observation than traditional methods.

1. The Common Implementation Strategy

Beyond the familiar legislative and comitology arrangements, responsibility for implementation is in principle with the Member States. Implementation is extraordinarily challenging, and the monitoring, identification, and enforcement of implementation equally so. The challenges of implementation have led to mechanisms for implementation beyond the letter of the legislation, in particular in an informal, but complex, demanding, and inevitably somewhat obscure joint approach to implementation. The Common Implementation Strategy provides for networks of experts to work together on the implementation of the Water Framework Directive.[80] These sorts of networks are a pervasive element of 'new' and of 'multi-level' governance. They are characterized by a lack of formal hierarchy and by cooperation between different levels of governance and between private and public entities.

The Common Implementation Strategy is organized at three levels: Technical Working/Drafting Groups, the Strategic Coordination Group, and the Water Directors.[81] The Technical Working/Drafting Groups prepare guidance and are composed of experts from any Member State that wishes to participate.[82] The Strategic Coordination Group is chaired by the Commission and made up of participants of the Member States. In addition, 'NGOs and stakeholders may be invited as observers and/or consulted',[83] and both the European Environment Bureau and the WWF have participated.[84] The Strategic Coordination Group is a political body responsible for coordinating the different working groups and activities. It evaluates the outcome of the different working groups and prepares

---

[80] Although the Directive does not refer to the Common Implementation Strategy, note also Recital 14: 'The success of this Directive relies on close cooperation and coherent action at Community, Member State and local level'.

[81] Although there is something of a proliferation of bodies, see Strategic Document, *Improving the comparability and the quality of Water Framework Directive implementation—Progress and Work Programme 2007–2009* (2006).

[82] Strategic Document, ibid.

[83] Strategic Document, ibid., p. 12.

[84] European Environment Bureau, 'Tips and Tricks' for Water Framework Directive Implementation (EEB, 2004).

documents and reports for the Water Directors. The Water Directors are the Member State representatives with overall responsibility for water policy, so senior civil servants. The Water Directors steer the process and take final responsibility for the published documents and for the strategy itself. Environmental interest groups have not been allowed to participate at this level.[85]

The activities of the Common Implementation Strategy include the exchange and management of information and data, as well as testing and validation of different approaches to the Directive. The Common Implementation Strategy allows for the pooling of good practice and information from diverse sources. It relies on learning and experimentation in a directive that attempts to measure and achieve things that have never been tried before. Effective and efficient implementation are clear objectives of the Common Implementation Strategy.[86] It allows Member States to share the no doubt significant technical and human resources required for implementation, increasingly important with greater economic disparities within an enlarged EU. And the Common Implementation Strategy is to contribute to enforcement by extension of peer review and by encouragement of good practice. Beyond the direct and hierarchical (and burdensome) recourse to Articles 226–228 of the Treaty, there have been gradual moves over recent years to enhance implementation of environmental law through education, assistance, and monitoring, culminating (for now) in activities like the Common Implementation Strategy. The Commission is 'convinced' that this leads to 'better results than a more formalistic approach to implementation'.[87] Whilst the Commission is keen to emphasize that the Treaty enforcement powers remain in the background, it is important also to remember that 'new governance' has been called on here in part precisely because of the limits of formal enforcement. By providing clearer background standards and information than the legislation itself, however, the Common Implementation Strategy may even make more traditional formal legal action more straightforward.

The most significant activity of the Common Implementation Strategy is the development of non-binding Guidance Documents.[88] The Guidance Documents are subject to testing and revision, presumably an effort to respond to heavy criticism of outdated and unscientific regulation of water in the past, as well as a

---

[85] Ibid.

[86] Strategic Document, *Common Strategy on the Implementation of the Water Framework Directive* (2001).

[87] Commission, supra n. 57, pp. 10–11. Note that the Commission refers to the Common Implementation Strategy in its annual report on implementation, see Commission Staff Working Document, *Seventh Annual Survey on the Implementation and Enforcement of Community Environmental Law 2005* SEC (2006) 1143, p. 8. This document contains a section on 'Non-legal instruments and initiatives to improve implementation', reminding us of the more general importance of this approach.

[88] The Common Implementation Strategy is most resonant of the IPPC Directive, supra n. 4 in which flexibility in legislation is subject to 'soft' harmonization beyond the legislation. See Scott, supra n. 4 for discussion.

more general recognition of the importance of learning in environmental regulation. Seventeen documents have been published up to the beginning of 2008, on subjects from analysis of pressures and impacts on water, to the identification of water bodies, but also on issues like monitoring, economic analysis, and public participation. These are very detailed, highly technical, and lengthy documents, but resting on profoundly political choices. The Guidance Documents set out agreed positions on areas in which costs, risks, and benefits are distributed, the bread and butter of political discourse, and have a major role in determining the environmental outcomes of the Water Framework Directive.

Whilst it is working in a range of areas of implementation, some of the most important work of the Common Implementation Strategy is in the further definition of the environmental objectives required in Article 4. The definition of ecological status is, as discussed above, profoundly uncertain. It requires reference to presumed pristine conditions, which means that expectations as to appropriate characteristics are legitimately very different for different bodies of water. This is a space for a flexibility that reflects ecological differences: the acceptable ecological parameters for a river in the Scottish highlands will be very different from a river in lowland Spain. But whilst based on principle, the flexibility makes it difficult to assess progress and to compare national implementation. Which is where the Common Implementation Strategy comes in. It will be recalled that ecological quality is assessed according to 'biological elements', 'hydromorphological elements supporting the biological elements' and 'chemical and physico-chemical elements supporting the biological elements'. And each type of water (rivers, lakes, coastal waters, etc.) is assessed against a range of characteristics within these broad categories. Member States are required to 'establish monitoring systems for the purpose of estimating the values of the biological quality elements specified for each surface water category'. To ensure comparability, the Member States have to express these results in the same way, 'as ecological quality ratios'. 'Ecological quality ratios' represent the relationship between the values of *actual* biological parameters and the values for these parameters in the reference conditions for such a body of water. We are to end up with a number between one and zero. This number will determine the categorization of the water body from high to poor, with high ecological status represented by values close to one and poor ecological status by values close to zero.

To ensure comparability, the value for the boundary between the classes of high and good status, and the value for the boundary between good and moderate status is to be established through an intercalibration exercise. The Common Implementation Strategy's role in intercalibration is to some degree foreseen by the Directive's assumption of collaboration and cooperation, but it is defined entirely beyond the legislation.[89] The Common Implementation Strategy allows

---

[89] 'Intercalibration' can be found in the part of Annex V of the Directive that deals with the 'comparability of *biological* monitoring results', para. 1.4.1. The Commission is to 'facilitate' an

for the negotiation, collaboration, and compromise that are necessary to calibrate biological parameters in the diverse water environments around the EU. The intercalibration exercise was supposed to have been completed in 2006,[90] pushed back to the end of 2007,[91] and is still (May 2008) delayed. This is a crucial and very difficult and controversial part of the Directive, and so difficulties gathering the data and reaching agreement are not surprising. But on the other hand, it is central to the operation of the Directive, and delay here is likely to lead to delay everywhere. Some of the results so far have been criticized by environmental interest groups who blame lack of data but also the failure of the Member States to devote the appropriate resources (especially experts) to the task.[92] The Commission accepts that there are 'deficiencies' in the intercalibration exercise,[93] and even the Common Implementation Strategy Working Group on intercalibration recognizes the problems, particularly the need to produce results on the basis of limited data.[94]

Ecological quality ratios and their intercalibration transform a multi-faceted and inherently flexible and varied idea of 'ecological status' into a numerical value. Intercalibration has significant potential value, but it is probably inevitable that efforts to put a number (or even a colour)[95] on something so complex will flatten our perspective. This is the cost of administrative workability and transparency, but it sits strangely with the integrated and holistic approach we see on the face of the legislation. Perhaps most interesting for current purposes, the enormous difficulty of intercalibration begins to answer questions as to the purpose of new governance in this Directive. The alternative to new governance was not clear and enforceable intercalibrated quality ratios in legislation, but either less ambitious objectives (less extensive biological criteria) or no comparability.

intercalibration exercise, 'to ensure that these class boundaries are established consistent with the normative definitions in [Annex V] and are comparable between Member States'.

[90] The final list register of sites was to be established within four years of the entry into force of the Directive, the intercalibration exercise completed within 18 months of that, and the results published within a further six months of that (Annex V, 1.4.1 (vii–ix)). The Directive entered into force on the day of its publication in the Official Journal (22 December 2000), Article 25. The register of sites was published in September 2005: see Commission Decision 2005/646/EC on the establishment of a register of sites to form the intercalibration network in accordance with Directive 2000/60/EC of the European Parliament and of the Council [2005] OJ L243/1. The list of sites is explicitly open to change on the basis of experience, Recital 6.

[91] Commission, supra n. 57.

[92] European Environmental NGO Technical Review of the Water Framework Directive Intercalibration Process (EEB, RSPB and Pond Conservation, 2006).

[93] Commission, supra n. 57, pp. 9–10.

[94] Common Implementation Strategy, Guidance Document No 6 *Towards a Guidance on the Establishment of the Intercalibration exercise* (2002): 'The WFD foresees a single intercalibration exercise in 2005 and 2006. It is inevitable that this exercise will be based on results from monitoring systems that are still under development, with limited data available and practically no possibility to collect additional data', p. 18.

[95] Bodies of water are to be presented on a map colour coded according to their ecological status (or potential) classification, Annex V, para. 1.4.2 (i)–(iii).

Even the poor progress so far seems to have reinforced the expectation that intercalibration will be an ongoing and learning process, 'a step-wise approach ... with improvements and refinements being introduced in the light of experience and new information', rather than a one-off.[96] Whilst not apparent on the terms of the Directive, if followed through, it is a practical approach that demands learning and progress. Far from being an inadequate substitute to legislation, if the pressure is kept on, the joint approach to measuring ecological quality allows legislation to attempt things it could not possibly have attempted otherwise. Again, what we learn about river basins seems to be as important as anything else.

2. Concerns about the Common Implementation Strategy

This chapter has looked at two different although closely related manifestations of new governance. One is the explicit use of particular tools and techniques on the face of legislation. The other is a whole world of implementation measures that could not be predicted at all from reading the legislation. Joanne Scott and Jane Holder go so far as to suggest that the experimentalism of the Water Framework Directive 'has been concealed by a legislative framework which rests on different, and increasingly misleading, premises'.[97] New governance undoubtedly raises some profound constitutional challenges,[98] but one relatively modest challenge must be the degree to which the extra-legislative governance mechanisms might distract or detract from legislative commitments. Whatever the potential environmental benefits of collaboration beyond the terms of the legislation, it is important to continue to interrogate the processes of governance against the claims of law in the Water Framework Directive.

Of course we will never know what environmental benefits might have been achieved through 'traditional' mechanisms. But the Water Framework Directive is ambitious and far reaching in its environmental objectives, notwithstanding their uncertainty and the significant exceptions and derogations. This ambition was arguably only possible because of the ability to negotiate detail through new governance. It is difficult, for example, to imagine that ecological quality ratios would ever have been 'intercalibrated' in a Directive. But although there are no guarantees to date that they will be through the Common Implementation Strategy, without the prospect of common implementation, ecological ambition was more likely to have been abandoned than subject to any clear legislation-implementation-enforcement chain. The consensual approach of the Common Implementation Strategy does nevertheless raise concerns about the ambition of the output.[99] The European Environment Bureau urges that the Guidance

---

[96] Common Implementation Strategy, supra, n. 94, p. i; see also p. 23.
[97] Scott and Holder, supra n. 2, p. 212.
[98] See the contributions to de Búrca and Scott, supra n. 2.
[99] European Environment Bureau, supra n. 84.

Documents be applied critically, and that the Commission assess implementation by reference to the Directive, not just the Guidance Documents.[100] But concerns about the extra-legislative imposition of unduly high environmental standards, closing down the legislative flexibility, are equally valid.

Put crudely, the most basic question of accountability is who to blame if implementation is inadequate. And the crude answer here has to be the Member State, which retains formal legal responsibility. But of course the definition of terms in the legislation is unclear, making sanctioning, and even blaming, problematic. At this point, the alternative tools in the Directive come into play. In particular obligations of publicity and explanation, as discussed above, are potentially powerful forms of accountability. And it is notorious that open information is not necessarily transparent[101]—highly technical information is not accessible to non-experts, and relevant information can be buried in a mass of data. So the Common Implementation Strategy can provide something special through peer review. Experts explain their approaches to other experts, national governments to other national governments. This is a form of transparency to people who really know how to use the information. It also provides some enhanced accountability *beyond* the Member State, and whilst there are always questions about the cost to accountability *within* the Member State, this is particularly appropriate in the EU context.[102] The Common Implementation Strategy contributes significantly to the accountability of Member States (to each other, to the Community, to certain carefully selected 'publics') when it would otherwise be too easy for them to escape accountability. It grapples with questions of control and responsibility to mediate the flexibility that has been deemed the best way to tackle a particular environmental problem. It provides central and peer influence to avoid backsliding, and a range of powerful tools to address Member State accountability in conditions of uncertainty.

This is enormously important, but it is all about accountability beyond the national arena. We would normally be equally concerned with accountability towards national publics. Administrative structures like the Common Implementation Strategy can sit uncomfortably alongside the strong commitment to transparency and participation in environmental governance, including in the Directive itself. To the extent that public scrutiny is designed to keep Member States on the straight and narrow in implementation, it could be argued that this is enhanced by the Common Implementation Strategy. But public participation also contributes perspectives that may not be identified by experts, as well as contributing a form of democratic legitimacy to a process. A need to respond to

---

[100] European Environment Bureau, ibid.
[101] See for example Jasanoff, S., 'Transparency in Public Science: Purposes, Reasons, Limits' (2006) 69 *Law and Contemporary Problems* 21.
[102] See for example Sabel, C.F. and Zeitlin, J., 'Learning from Difference: The New Architecture of Experimentalist Governance in the European Union', *European Governance Papers* C-97-02 and their discussion of 'democratizing destabilisation'.

peer scrutiny could divert attention from the need to respond to the demands of the public.[103] Whilst the Common Implementation Strategy could enhance the accountability of the Member States in respect of the Directive, the nature of the Common Implementation Strategy itself remains an afterthought. There is relatively little apparent concern with the legitimacy of 're-Europeanisation' through new governance, beyond the letter of the law, and more cynically beyond the view of the sceptical European publics.

Who participates in the Common Implementation Strategy is an important question. Guidance Documents are public documents, and in a welcome piece of transparency, the members of working groups (although not always their affiliations)[104] are listed. Apparently over 1,000 experts have been involved in the networking possibilities of the Common Implementation Strategy,[105] and divided and different interests are present. But the proliferation of deliberative or negotiation networks through Common Implementation is beginning to sound enormously burdensome. The Water Directors have acknowledged criticism that 'the work intensive process may put additional pressure on the national authorities', already faced with a high work load on this Directive.[106] This is a major practical challenge and might be tackled by prioritizing resources, or even ultimately by abandoning parts of the process. It is a deeper problem when one turns to the question of equality of access, since it seems likely that these processes will be dominated by the larger, wealthier Member States.[107] Large and well organized industry will also generally find the resources to take part when their interests may be directly at stake.[108] But routine involvement of environmental interest groups across Common Implementation Strategy activity is highly vulnerable, and at best generally restricted to large well resourced groups. Poor representation of environmental interest groups could allow networks to simply turn into a complex negotiation between industry and regulators.[109]

Given the very specific interests of different participants, and the likelihood that even at their best they do not capture the full range of issues around the regulation of water, what actually goes on in these meetings becomes very important. There is a danger that the accountability functions of the Common

---

[103] See Papadopoulous, Y., 'Problems of Democratic Accountability in Network and Multilevel Governance' (2007) 13 *European Law Journal* 469.

[104] When affiliations are given, acronyms seem to rule, and identifying which interests have had a say in drafting is not easy.

[105] Strategic Document, supra n. 81.

[106] Strategic Document, ibid.

[107] This does seem to have been the case for IPPC, see European Commission, *On the Road to Sustainable Production: Progress in Implementing Council Directive 96/61/EC Concerning Integrated Pollution Prevention and Control* COM (2003) 354 final.

[108] The Commission seems to acknowledge that industry's access to resources of expertise and information gives it a certain influence in the IPPC process. In particular, the 'most comprehensive information' on BAT comes from industry, Commission, ibid., p. 17.

[109] Hey, C., *Towards Balancing Participation* (European Environmental Bureau, 2000), discussing IPPC.

Implementation Strategy could be blunted by mutual interest, as groups of expert peers (consciously, or more likely without sinister intent), collude along lines that fit with their own understandings of the problem.[110] Or these meetings may provide a space for deliberation in the public interest, or they may be 'business as usual' interest representation and compromise. Deliberative forms of democracy have had a great appeal in the context of governance of environmental problems at EU level. The potential for the transformation of values and ideals, the notion that decisions are properly based on values, supported by reasons, rather than on competition between self-interested positions, is instinctively attractive. Whilst there is currently insufficient evidence to reach a conclusion on deliberation versus negotiation in the Common Implementation Strategy, it would perhaps be surprising if we did not see both public spirited deliberation and self-interested negotiation at different moments.[111] But what counts as public spirited is precisely what is at stake here, both in the appropriate balance between environmental and economic objectives, and in quite what environmental protection demands in particular cases.

To the extent that the Guidance Documents produced through the Common Implementation Strategy are non-binding, subject to application if at all by regulators within the Member States, this may seem unimportant. Indeed, the Common Implementation Strategy emphasizes the 'fundamental principle' that responsibility for implementation is 'fully within the competence of the individual Member State'.[112] However, the Guidance Documents are potentially enormously influential in a very demanding implementation process, particularly in Member States with fewer resources to devote to environmental regulation (precisely those Member States least likely to be influential in the negotiation of the documents).[113] The Guidance Documents carry considerable scientific and political, if not legal, authority, and it is likely that Member States will at least feel constrained to explain a failure to comply with them. Indeed, in an indication of the blurred boundaries between governance techniques and 'hard' law, the Commission has proposed legislation that would require such an explanation in respect of somewhat similar 'Reference notes' that

---

[110] Harlow, C. and Rawlings, R., 'Promoting Accountability in Multi-level Governance: A Network Approach' (2007) 13 *European Law Journal* 542.

[111] Bettina Lange's study of IPPC suggests that interest representation (and hence negotiation and compromise, rather than simple fact finding) can be expected in the committees, Lange, B., 'From Boundary Drawing to Transitions: The Creation of Normativity under the EU Directive on Integrated Pollution Prevention and Control' (2002) 8 *European Law Journal* 246.

[112] Strategic Document, supra n. 81.

[113] Intriguingly, it seems to be the case that the reports required under Arts 3 (on river basin districts and competent authorities) and 5 (analysis) there was 'only limited or no use made of the CIS Guidance Documents', ibid., p. 2. But it is early days, and if the Common Implementation Strategy documents are not ultimately sufficiently influential to merit debate about the method of their elaboration, then the enormous resources devoted to it look rather misplaced.

are negotiated during the implementation of the Integrated Pollution Prevention and Control Directive.[114]

But notwithstanding the authority of the Guidance Documents, real local participation depends upon the ability of the consulting body to take its own decisions. And the Guidance Documents, published only in English, are highly technical and lengthy. They are likely to be read and critiqued only by competing experts, rather than the public more generally. It looks increasingly difficult to get social issues (such as the role of the River Alt, above, as social barrier) onto the agenda after the Common Implementation Strategy has spoken on an issue. Arguably, the Water Framework Directive has not rejected centralized substantive standards, but simply shifted the forum for their negotiation from legislation to networks, or using the language of old government, committees. Local participants can challenge the appropriateness of non-binding guidance in general or in the particular local circumstances.

There is potential for the two different fora of new governance, with their different preoccupations and concerns, to be occasionally at odds. But the main challenge that the Water Framework Directive seems to throw up in this respect is the long-standing and very familiar dilemma about the appropriate relationship between public and expert in decision-making in technical areas. It is a tension apparent throughout the Guidance Document on public participation,[115] and is compounded by the 'multi-level' nature of the governance problem, which creates the appearance of the experts being included at EU level, and the public being excluded at the local level. This is a dilemma that new governance shares with environmental policy far more generally and over many decades. The added difficulty is that the shadowy nature of implementation allows the mismatch between participation at central and local level (different constituencies, and a different scope of debate) to be fudged. Flexibility and non-hierarchical governance mechanisms do not necessarily imply greater or wider participation,[116] and can work to exclude as well as include.

The increasingly generalized perception that some sort of move away from the 'Community method', and especially something beyond comitology, is needed for effective and accountable EU governance has freed the governance debate to some extent. It is no longer forced on every occasion to respond to a mythical (especially in the EU) ideal of straightforward lines of accountability in legislation. Whilst conscious of the shift from well established mechanisms of legislative

---

[114] See European Commission, *Proposal for a Directive of the European Parliament and of the Council on industrial emissions (integrated pollution prevention and control)* COM (2007) 844, Articles 14–16. There are also proposals to convert certain guidance documents into legislative standards, ibid. The Commission has long said that it sees the 'BAT reference notes' in IPPC as the starting point for enforcement, see Emmott, N., Bar, S., and Kraemer, R. A., 'Policy Review: IPPC and the Sevilla Process', *European Environment* 10 (2000) 204, and that may now become more formalized.

[115] Supra n. 75.

[116] Smismans, S., 'New Modes of Governance and the Participatory Myth' (2006) *European Governance Papers* 06–01.

legitimacy,[117] the new governance framework can begin to scrutinize its performance in its own terms. And whilst the performance of the Common Implementation Strategy in this respect is far from ideal, minimum expectations on transparency are being developed. The pro-forma rhetoric of transparency and participation found in the Common Implementation Strategy documents is not adequate. But despite being an administrative structure on which no explicit demands are made, the Strategy has chosen to make itself highly visible and to allow scrutiny by groups with the expertise to make that scrutiny meaningful.

5. Conclusions

The Water Framework Directive takes an ambitious stand on the challenges of the EU's water environment. Its relative ecological holism, and its very demanding reference points are innovative and have great potential. The Water Framework Directive is aiming for things that have not been attempted before, and its ambition requires flexibility, in order to recognize both ecological and social realities. Because flexibility is an integral part of ecological thinking (especially in its recognition of diversity), but also a necessary partner to complexity and difficulty. The flexibility is constrained, first, by the explicit terms of the Directive, which without commanding particular results, disrupt normal ways of doing things by demanding particular types of information, assessments, and external scrutiny. The flexibility is further constrained beyond the terms of the legislation through the Common Implementation Strategy, which enhances accountability between Member States and between Member States and the centre, and to certain expert outsiders.

The implementation of the Water Framework Directive is still at an early stage, and this enigmatic Directive will continue to teach us a great deal about the role of new governance over the next decade and more. The environmental, economic, and social consequences of the Directive are obviously central to our assessment of the Directive's implementation. Currently, prospects for good status of waters by 2015 are not good: the Commission's first implementation report identifies a very high proportion of waters at risk of not meeting the good status standard.[118] The ultimate level of good status water will be one significant factor in judging the effectiveness of new governance techniques. What *counts* as good water status will be equally significant.

---

[117] See for example Shapiro, M., 'Administrative Law Unbounded: Reflections on Government and Governance' [2001] *Indiana Journal of Global Legal Studies* 369; Allott, P., 'European Governance and the Re-Branding of Democracy' (2002) 27 *European Law Review* 60.

[118] Commission, supra n. 57: 'Current status of EU waters—worse than expected', p. 3. This seems to be in part because of failure to implement earlier directives.

3

REACH: Combining Harmonization and Dynamism in the Regulation of Chemicals

*Joanne Scott**

1. Introduction

Europe has a new Chemicals Agency. Its establishment is inextricably tied to the emergence in Europe of an ambitious new framework for the regulation of chemicals. This framework is known by the acronym REACH, and it concerns the Registration, Evaluation, Authorisation and Restriction of Chemical Substances.[1] The establishment of a *European* Agency, and the European-level *harmonization* of chemicals policy seems to mark an important step in the direction of a centralization of regulatory power. However, it is the argument of this chapter that reality is more complex, more subtle, and certainly more promising. The European Agency does not stand on high at the apex of a hierarchically organized system for the regulation of chemicals. It forms part of a system of governance which is intensely fractured. Power is shared among a multiplicity of actors, operating at different levels of government, and in the private as well as the public spheres. No single actor has autonomous decision-making power. Rather, each is empowered, in different circumstances and in different ways, to play a role in maintaining the dynamic quality of regulation in the face of information deficits and uncertain risks. Different actors can play a role in seeking to ensure the continuous generation of new and better information about risk and about the mitigation of risk, and in seeking to prompt regulatory decisions which are appropriately responsive to this. This fracturing of power creates a governance framework which is complex. But it is also a framework which seeks to combine,

---

* Many thanks to John Applegate, Maria Lee, and Peter Oliver for very useful comments on an earlier draft. The students in the EUI Academy (2007) were outstanding in helping me get to grips with the intricacies of REACH. Many thanks to them also.
[1] Regulation 1907/2006. This both lays down the REACH framework and establishes the European Chemicals Agency.

in a novel way, harmonization with dynamism, and uniformity with structures for regulatory learning.

In the year 2000 the Lisbon European Council framed an ambitious goal for Europe; in short, to become the most competitive and dynamic knowledge-based economy in the world. The chemicals sector is one in which Europe has achieved considerable economic success. Nearly half of the world's leading chemicals companies are based in Europe, and Europe enjoys a healthy and growing trade surplus in chemicals.[2] Nonetheless, until recently at least, Europe's chemicals regulation was not worthy of this success, representing an impediment to innovation, and an insufficient basis for sustainable growth. The reasons for this were many. But the pernicious bottom line was the incentive structure generated by the regulatory framework. There had grown up a regulatory gap in the treatment of old and new chemicals.[3] The regulatory burden attaching to new chemicals entering the market was such that it was easier and less costly to continue to use old chemicals which were largely exempt. The impetus to innovation was dulled and industry convenience came to prevail over safety. The EU's new chemicals regime represents an ambitious attempt to alter the incentive structure between old *and* new chemicals, and to do so in a manner which achieves a high level of protection of human health and the environment.

The volume and complexity of the Regulation is such that a lengthy book would be required to offer a full exposition and analysis. Running to 141 articles, 17 annexes, and 10 appendices, the densely printed Official Journal text amounts to 280 pages. At the heart of these 280 pages is a system for the regulation of old and new chemicals which makes the generation and dissemination of information about chemicals a priority. This information is intended to ensure that sensible decisions can be taken about which substances to focus investigative attention on, about when and how to restrict the use of chemicals, and about when to ban them completely. It is also intended to ensure that market transparency will create conditions in which chemical users—commercial users as well as consumers—will not only be properly informed about safe use, but will be able to exercise choices in favour of safer alternatives. When it comes to high-risk categories of chemicals, the availability of suitable and safer alternatives, will be a reason for the regulator to ban them.

In keeping with the theme of this volume as a whole, this contribution focuses upon the governance approach embodied in the REACH regulation, and upon

---

[2] For facts and figures on the European chemical industry see <http://www.cefic.be/factsandfigures/level02/profile_index.html>.

[3] For this purpose old chemicals are known as existing chemicals and are those dating from before 1981. See Koch, L. and Ashford, N.A., 'Rethinking the role of information in chemicals policy: implications for TSCA and REACH', *Journal of Cleaner Production* 14 (2006) 31 at <http://web.mit.edu/ctpid/www/tl/docs/Koch&Ashford1.pdf> for a brief overview and critique of the EU's chemical regime. See also for an excellent outline of the main elements and legal issues arising, Molyneux, C.G., 'Chemicals' (2008), 8 *Yearbook of European Environmental Law* (2008).

certain key features ('core elements') which characterize this. The core elements pinpointed in this paper are industry responsibility, contestability, substitution, provisionality, and transparency. Three observations, flowing from the discussion which follows, may be usefully highlighted from the start.

First, and perhaps most important, REACH is concerned with product regulation, the regulation of chemical substances. Product regulation here, as elsewhere, bears heavily on trade, in both the EU and the WTO. This trade dimension comes to the fore in the regulation. It is exhaustive in the harmonization which it achieves. It takes Article 95 as its legal basis, and contains a clear free movement clause. Member State recourse to the treaty-based free movement exceptions will be pre-empted. By contrast to much recent environmental law in the EU, the regulation is about harmonization not diversity, and offers reduced flexibility to Member States. But this surface impression does not tell the whole story. This is because the harmonization which it achieves is provisional and contestable. A wide range of mechanisms are established to allow for the contours of the provisional harmonization bargain to be contested, including in the light of experience in the individual Member States. At every turn, whether it is in relation to which substances are to be evaluated, which are to require authorization or restriction, or in relation to labelling requirements, Member States (and the Commission) are empowered, on the basis of clearly defined procedures, to seek to use their local knowledge to persuade the European Union as a whole, of the need to revise applicable norms, in order to ensure effective fulfilment of the Regulation's framework goals (high level of protection of human health and environment in particular). By contrast to the more traditional Article 95 EC Treaty opt-out, these procedures for contestation are multi-level and multi-actor, and eschew an approach which places the Commission on high, supposedly omniscient in its capacities to make judgements about regulatory right and wrong. Because of this, REACH offers an important lesson in thinking about governance in EU environmental law. By making harmonization provisional and contestable, and by creating mechanisms for contestation and regulatory learning, the EU has begun to articulate a model for regulatory federalism which seeks to respond to the imperatives of free trade *and* those of environmental/health protection.

Second, although in the form of a regulation, the implementation phase is all-important under REACH.[4] This becomes starkly clear by looking at the list of substances requiring authorization; empty at the time of enactment! The governance arrangements for the adoption of implementing decisions are varied and

---

[4] As John Applegate points out, it was at the implementation stage that the US chemicals regime in the form of the Toxic Substances Control Act, met its most pronounced failures. Careful scrutiny of implementation practices and decisions will be essential to any long-term evaluation of REACH. See Applegate, J. S., 'Synthesizing TCSA and REACH: Practical Principles for Chemical Regulation Reform', *Ecology Law Quarterly* forthcoming.

complex. As is typical, comitology procedures play an important part.⁵ But on the whole the overall framework for decision-making is more inclusive than in the past.⁶ This is not only a result of the enhanced role *sometimes* played by the European Parliament as a result of the new regulatory committee with scrutiny procedure.⁷ It is also because interested third parties are given significant opportunities to participate in decision-making processes, and to have their viewpoints taken into account. In the discussion of substitution below, it is suggested that these opportunities for participation create a mildly adversarial mechanism, somewhat reminiscent of the kind that Wendy Wagner has in mind in her elaboration of a competition-based approach to the regulation of chemicals.⁸

Third, REACH places considerable emphasis upon information as a regulatory tool. At its heart, it is intended to mitigate the overwhelming data deficit which exists in relation to older chemical substances. REACH is also concerned with the communication of information, up and down the supply chain, and to consumers and to the public at large. As it stands, however, there exists an important gap. Labelling requirements relate only to chemical substances and preparations (mixtures of chemical substances). Consumer articles, such as textiles or furniture for example, containing chemical substances are not covered. Even where an article contains a chemical substance of very high concern (for example a carcinogenic substance) consumers will not be alerted to this by the presence of a label. Thus, to the extent that REACH seeks to harness the power of the marketplace to encourage innovation in the development of less risky alternatives, its labelling rules would seem to fall short.

The discussion which follows is often technical and dense. This is unavoidable when examining a regulatory regime of this volume and complexity. This should not be allowed to disguise either the intrinsic importance of the subject under discussion for environment and human health protection,⁹ or the fascination of the governance forms which this regulation embodies. It is in the minutiae of law's construction of decision-making procedures and conditions for action that the politics of risk regulation are played out.

---

⁵ For background to the concept of comitology see Craig, P. and de Búrca, G., *EU Law: Text, Cases and Materials* (OUP, 2007), pp. 118–23.
⁶ See on this theme LSE Working Paper 07–2008, Heyvaert, V., 'The EU's Chemicals Policy: Towards Inclusive Governance?' at <http://www.lse.ac.uk/collections/law/wps/WPS2008-07_Heyvaert.pdf>.
⁷ See again Craig and de Búrca, supra n. 5. For further details of the new regulatory committee with scrutiny procedure, see Article 5a Council Decision 1999/468. By 'sometimes', it is simply meant that this new procedure applies to some decisions to be adopted under REACH, but by no means all. Further details on this will emerge in the sections which follow.
⁸ Wagner, W., 'Using Competition-Based Regulation to Bridge the Toxics Data Gap' at <http://search.ssrn.com/sol3/papers.cfm?abstract_id=1090090>.
⁹ See, for a brief overview, chapter 1 of the Royal Society for Environmental Pollutions chemicals study at <http://www.rcep.org.uk/chreport.htm>.

2. Core Element: Industry Responsibility

A. Registration

1. What does registration require?

REACH imposes an informational burden on industry. It does so through the introduction of a default rule: 'no data, no market',[10] generating a strong incentive for the information to be provided. Chemical substances manufactured or imported at or above a one tonne threshold must be registered with the new Chemicals Agency.[11] The Regulation establishes detailed rules for identifying substances requiring registration.[12] These include substances on their own, or in preparations (such as a detergent).[13] Also included are substances in articles, where the one tonne threshold is met, and where the substance is intended to be released under normal or reasonably foreseeable conditions of use.[14] Even where there is no intention that the substance be released, the Agency may require registration of substances in articles where it has grounds for suspecting that the substance is released and that this presents a risk to human health or the environment.[15] A number of substances enjoy total or partial exemptions under the regulation.[16]

---

[10] Article 5.

[11] Article 6(1). The threshold is determined at the level of an individual manufacturer or importer, and calculated on an annual basis. Note that in some situations a notification obligation will apply even in the absence of a registration requirement. This implies a much reduced informational burden but an informational burden all the same. See especially Article 9(2) in relation to the R & D exemption. See also n. 14 below in relation to notification of substances in articles under Article 7(2). Note also that the reference to 'notified substances' in Article 24 refers to a different concept. Here, notification under the old regime is being treated as registration for the purpose of the new one. On the Agency see Articles 75–111 and: <http://echa.europa.eu/>.

[12] The general definitions are provided in Article 3 where the concept of a substance is defined as a chemical element and its compounds in the natural state or obtained by any manufacturing process, including any additive necessary to preserve its stability and any impurity deriving from the process used, but excluding any solvent which may be separated without affecting the stability of the substance or changing its composition.

[13] Article 6(1). A preparation is defined in Article 3(2) as a mixture or solution composed of two or more substances.

[14] An article is defined in Article 3(3) as an object which during production is given a special shape, surface, or design which determines its function to a greater degree than does its chemical composition. Note a reduced obligation to notify (as opposed to register) the Agency of substances of very high concern in articles, where it is present above a concentration of 0.1% w/w, even where it is not intended to be released. The concept of substances of very high concern will be raised in the discussion below on authorization. This lesser notification obligation can only be excluded where the producer or importer can exclude exposure to humans or to the environment during normal or reasonably foreseeable conditions of use. See Articles 7(2) and (3). Registration would thus seem to rest on a subjective test (intended to be released), whereas notification rests upon an objective test.

[15] Article 7(5). This is still subject to the one tonne threshold. The Agency's decision is not subject to appeal by the Agency Board of Appeal (see Article 91) but it is susceptible to review by the Court of First Instance (see Article 94).

[16] See in particular Articles 2 and Annexes IV and V. Included among other things in Annex V are substances occurring in nature if they are not chemically modified or dangerous.

Prominent among these are polymers,[17] and certain intermediaries.[18] Likewise, substances manufactured or imported for the purposes of product or process oriented research and development are automatically exempted for five years.[19] This may be extended, upon request, by the Agency for a further five-year period; ten years in the case of medicinal products for human or veterinary use.[20]

The registration requirement took effect for new substances on 1 June 2007. Existing substances, known as phase-in substances, are able to benefit from transitional arrangements.[21] For these substances registration will proceed in stages, beginning in 2010 for very high volume substances and for certain risky substances, and ending in 2018.[22] Pre-registration is required for phase-in substances seeking to benefit from these transitional arrangements.[23] Upon submission of a registration document, the Agency will perform a completeness check. Where a registrant receives no indication to the contrary from the Agency within a three-week period, it may begin or continue to manufacture or import the substance in question.[24]

The intensity of the informational burden imposed by REACH is variable. First, and critically, the basic requirement rests upon an open-ended standard, rather than a detailed rule. This requires that each registrant must submit a technical dossier, and this must include all physico-chemical, toxicological, and

---

[17] These are exempted from registration, although requirements may be extended in the future to polymers where a practicable and cost-effective way of identifying risky polymers is established. Polymers are large molecules consisting of repeated chemical units joined together. Plastics are polymers, for example.

[18] Intermediaries are chemicals used to make other chemical substances. On-site and transported isolated intermediaries are subject to limited registration requirements subject to the substance being rigorously contained during its whole life-cycle. Monomers used for this purpose do not fall within this. See Articles 17–19.

[19] Article 9.

[20] See Article 9(7) for the conditions which apply. Any decision taken by the Agency here will be subject to appeal.

[21] Article 23. See Article 3(2) for a confusing definition of the concept of phase-in substances. These include substances included in the European Inventory of Existing Chemical Substances (EINECS). Note that where an existing substance has been notified under the earlier regime under Directive 67/548, this will be regarded as a registration for the purpose of REACH. See Article 24. The aim of the regulation is to address the 'burden of the past' meaning substances not covered by the earlier regime.

[22] Very high volume means 1,000 tonnes or more here, and the risky substances to be registered by 2010 are those which are category 1 carcinogens, mutagens, or reproductive toxicants (CMRs) (no volume threshold) and those which are very toxic to the aquatic environment posing risks of long-term adverse effects (100 tonne threshold). See Article 23(1).

[23] Article 28. The relevance of this will become clear when talking about joint submission and data-sharing below.

[24] See Article 20(2) on the completeness check and Article 21(1) on timing. Note that the Agency has a three-month rather than a three-week period to respond in relation to phase-in substances registered within two months of the end of the transitional period. This is presumably due to concerns about the high number of last-minute registrations which will be submitted at this time, and reflects the fact that there are considerably more existing than new substances.

ecotoxicological information that is *relevant* and *available* to the registrant.[25] This standard is accompanied by more specific rules setting out the minimum information to be included. In the main, the nature of this minimum will be determined by reference to volume.[26] For registered substances, more information is demanded as production volume increases. For the lowest tonnage level (up to 10 tonnes) the standard requirements in Annex VII will apply.[27] Every time a new tonnage threshold is reached (10 tonnes, 100 tonnes, and 1,000 tonnes) an additional annex will bite, imposing additional informational demands. Information on the intrinsic properties of substances may be generated by means other than tests, provided that the conditions set out in Annex XI are met.[28] Indeed, wherever possible, information shall be generated by means other than vertebrate animal tests,[29] and all available *in vitro* data, *in vivo* data, historical human data, data from valid qualitative or quantitative structure-agency relationships ((Q)SAR), and from structurally related substances (read-across approach) is to be assessed first before new tests are carried out.[30] But as noted, such data will only be able to substitute for the results of the standard testing regime where they meet the requirements of Annex XI. These are expressed in open-ended, standard-based, terms; adequacy, reliability, and scientific validity for example. It will be for the Agency to evaluate registration documents to ensure that these requirements have been met.[31]

Two additional points may be usefully made in relation to the nature of the informational burden incurred by industry. First, Annex IX and X apply at the 100 tonne and 1,000 tonne marks respectively. In each case, the registration is to include testing proposals for the provision of the specified information, rather than the information itself; subject to the broad disclosure standard set out above.[32] It will be for the Agency to examine testing proposals, and to draft a decision accepting, rejecting, modifying, or supplementing the testing proposals.[33] Second, the 10 tonne volume threshold is of particular pertinence.

---

[25] Article 12(1). A guidance note on information requirements is set out in Annex VI, but does little to build upon this. It does make it clear that registrants have to be somewhat proactive in collating information by, for example, undertaking a literature search.

[26] There is one exception to this. More information has to be submitted for Annex III phase-in substances than non-Annex III phase-in substances. For the latter no toxicological or ecotoxicological information is required, but only physico-chemical information in accordance with s. 7 of Annex VII. Annex III substances include category 1 and 2 CMRs, as well as substances with dispersive or diffuse effects for which it is predicted that they are likely to meet the classification criteria for any human health or environmental effects endpoints under Directive 67/548.

[27] Subject to the previous footnote for non-Annex III phase-in substances.

[28] Article 13(1).   [29] Ibid.

[30] Annex VII, penultimate Recital.

[31] It will do so in the course of dossier evaluation to be discussed below. See especially Article 41(1)(b) and the rubric to Annex XI.

[32] See Article 12(d) and (e) and Annexes IX and X.

[33] Article 40(3). Where any Member State proposes an amendment to the Agency draft decision, and where the Member State committee is not unanimous in its support, the final decision will be adopted by the Commission on the basis of a regulatory committee procedure.

It is at this stage that the obligation to conduct a chemical safety assessment and to submit a chemical safety report (CSR) kicks in.[34] This will require human health, environment, and physico-chemical hazard assessments, as well as an assessment of whether the substance may be characterized as persistent, bioaccumulative and toxic, or very persistent and very bioacumulative (PBT/vPvB). For dangerous substances and those which are PBT/vPvB, an enhanced CSR will be required, to include an exposure assessment and risk characterization.[35]

2. Industry accountability in registration

It has been suggested that REACH creates perverse incentives for industry. '[P]roducers have an incentive to underestimate risk in order to avoid outside intervention into self-regulation. This seems to be a key governance problem of REACH.'[36] It is then essential to consider the range of mechanisms put in place in an attempt to keep industry honest and forthcoming in its provision of information. Two principal mechanisms are put in place; one premised upon hierarchical oversight, and the other upon peer review.

Hierarchical control is largely predicated upon the concept of dossier evaluation.[37] This involves two core elements, an examination of testing proposals and a compliance check of registrations.[38] Where, on the basis of compliance checking, the Agency concludes that the registration is not in line with the relevant information requirements, it may prepare a draft decision requiring the registrant to submit additional information.[39] For compliance checking, the Agency is required to examine at least 5% of registrations in each tonnage band.[40] It shall give priority, 'but not exclusively', to certain categories of registrations, namely those falling short in providing the standard Annex VII information, those concerning priority substances included on the Agency rolling action plan for substance evaluation, and those in respect of which specified information has

---

[34] See Article 14 and also Annex I containing general provisions for assessing substances and preparing chemical safety reports. Annex I, unlike the previously mentioned Annex IV, is not framed as a guidance note.

[35] Article 14(4) and Annex I, sections 5 and 6.

[36] Hey, C., Jacob, K., and Vokery, A., 'Better Regulation by New Governance Hybrids: Governance Models and the Reform of European Chemical Policy', Environmental Policy Research Centre paper at <http://papers.ssrn.com/sol3/papers.cfm?abstract_id=926980> p. 13.

[37] Title VI, chapter 1.

[38] Compliance checking should be contrasted to the previously mentioned concept of a completeness check under Article 20. By contrast to compliance checking, the completeness check 'shall not include an assessment of the quality or the adequacy of any data or justifications submitted', and is more a tick-box exercise about the categories of information received.

[39] Article 41(3). The procedure for the final adoption of this decision is laid down in Articles 50 and 51. In essence, the Agency enjoys autonomy unless it receives a Member State proposal for the amendment of its draft decision. In this event, the Agency can only decide if it enjoys the unanimous support of the Member State committee, otherwise the Commission will decide acting on the basis of a regulatory committee procedure.

[40] Article 41(5). This percentage may be varied on the basis of a regulatory committee with scrutiny procedure. See Article 41(7).

been submitted separately rather than jointly by manufacturers and/or importers.[41] Also of practical assistance in assisting the Agency in concentrating its compliance checking efforts, is the fact that any technical dossier submitted upon registration must indicate, for certain key categories of information, where this has been subject to review by an assessor prior to submission.[42] While there is no requirement that any such assessor be independent, it is stated that he or she must have appropriate experience. It may well be in the interests of companies to flag up the existence of internal procedures for review, in a bid to obviate the need for time-consuming and costly external review.

One of the stated grounds for prioritizing registrations for compliance checking hints at the nature of the second, non-hierarchical, mechanism for promoting industry accountability in the provision of information. The regulation establishes a binding, but non-absolute, requirement that key information, including physico-chemical, toxicological, and ecotoxicological information be jointly submitted by manufacturers and/or importers, with one registrant acting as lead registrant for this purpose.[43] A registrant may offer a reasoned justification for separate submission of this information on specified grounds only. These are disproportionate cost, substantial commercial detriment caused by the commercially sensitive nature of the information, and disagreement with the lead registrant on the selection of information.[44]

The mechanics of joint registration are different in the case of non phase-in (new) and phase-in (existing) substances. For non phase-in substances, potential registrants have a duty to inquire from the Agency whether a registration has already been submitted for the same substance.[45] If several potential registrants have made an inquiry for the same substance, the Agency shall inform each about the other, thus facilitating joint registration. For phase-in substances seeking to benefit from the transitional arrangements, there is a duty to pre-register, and to participate in a substance information exchange forum (SIEF).[46] Among other things, this SIEF will facilitate the exchange of key information including the physico-chemical, toxicological, and ecotoxicological information to be jointly submitted.

---

[41] Article 41(5).

[42] This relates to physico-chemical, toxicological, and ecotoxicological information under Annexes VII–XI, as well as chemical safety reports and information on manufacture, use, and classification and labelling of a substance.

[43] See Article 11 referring to categories of information in Article 10. These categories of information are, in turn, spelt out in greater detail in Article 12. Certain additional categories of information *may* be jointly submitted at the discretion of the manufacturers/importers, including the chemical safety report, while others *must* be submitted separately, including specified information on manufacture, use, and exposure.

[44] Article 11(3).

[45] Article 26. This also applies to phase-in substances which are not making use of the Article 23 transitional arrangements.

[46] Articles 28 and 29.

It is possible to speculate that the concept of joint registration might promote industry accountability in the provision of information, by inculcating an ethos of peer review. Where a manufacturer or importer disagrees with the information to be submitted, that manufacturer or importer may justify remaining outside of the registration consortium. Any attempt on the part of a registrant to manipulate data, or to be less than forthcoming in sharing relevant (unreliable) data, runs the risk of being signalled to the Agency by an absence of cooperation between industry actors in registration. As noted, departure from the default position of joint registration serves to increase the likelihood of hierarchical, Agency, compliance control; the fact of individual submission being one of the factors according to which the Agency will concentrate its efforts in compliance checking. Needless to say, the accountability reinforcing potential of joint submission is, even at best, less than absolute. It would do nothing to prevent industry consortia from collectively withholding or manipulating data.

Along with joint registration, data sharing also emerges as a key feature of REACH.[47] This is intended in large part to avoid and reduce vertebrate animal testing.[48] It is interesting to ask whether it might serve also to facilitate industry accountability in the provision of information.

For registered substances, information submitted as part of registration will become freely available after twelve years.[49] Even during this period, previous registrants have an obligation to share information with potential registrants in so far as this information involves testing on vertebrate animals.[50] Even where a substance has not been previously registered, all potential registrants participating in a SIEF are obliged to request and share data involving testing on vertebrate animals.[51] Again, subject to rules on cost-sharing, any holder of such a study will be precluded from proceeding to registration until he provides the study in question.[52] If a relevant study involving tests (all tests, and not merely tests on vertebrate animals) is not available, only one such study will be conducted per

---

[47] See Title III, Articles 25–30. This will be facilitated by initiatives not formally part of the REACH framework. See for example the launch by OECD of eChemPortal: The Global Portal to Information on Chemical Substances. This Internet gateway provides free access to information on intrinsic properties of chemicals as well as hazard and risk assessments. It allows for the simultaneous search of multiple databases.

[48] See Article 13 and Annex XI establishing the range of data sources which can be used to provide information on intrinsic properties on substances, and adumbrating the circumstances in which this can be used.

[49] See Articles 26(3) and 27(1).

[50] Article 27(1). Previous registrants may choose to share other categories of information. Article 27 lays down the procedure to be followed where there is no agreement between previous and potential registrants. Subject to rules on cost-sharing, ultimately the Agency will disclose the relevant information. The Agency's decision will be subject to appeal.

[51] Article 30.

[52] Article 30(3).

information requirement within the SIEF for the benefit of all members, and the resulting information duly shared.[53]

While in some respects the regulation does anticipate disagreements between registrants (actual and potential) on data sharing,[54] it does not contemplate circumstances in which one registrant remains dissatisfied with the content or methodology of a study conducted by another registrant. Not only does it not impose an obligation on a registrant to repeat the study in these circumstances, but neither does it expressly permit it to do so. Yet from the point of view of industry accountability it is critical that received wisdom be susceptible to challenge. In view of this, the underlying premise that testing on vertebrate animals be undertaken only as a last resort, and that the duplication of other tests should be limited, must be read as leaving room for additional testing where the reliability of existing data has been credibly questioned. The concept of data sharing is an important one. It has the potential to strengthen arrangements for peer review between industry actors by increasing transparency and mutual oversight. At the same time peer review cannot function in a meaningful manner if earlier tests, and the results emanating from them, are treated as anything more than presumptively valid.

B. Applications for Authorization

Certain substances of 'very high concern' may not be placed on the market without authorization, even at very low volume, and below the one tonne threshold for registration.[55] These substances are to be listed in Annex XIV.[56] When seeking authorization, applicants must 'analyse the availability of alternatives and consider their risks, and the technical and economic feasibility of substitution'.[57] Applicants must likewise submit the specified information, including a chemical safety report where this has not been previously submitted as part of the registration process.[58]

Like registration, authorization imposes an informational burden on industry. But it does more than this. In most cases authorization will be granted where the risk to human health or the environment is adequately controlled. It will be for the applicant to document this in its CSR.[59] This will require the applicant to

---

[53] Article 30(2). The members of the SIEF shall take reasonable steps to reach agreement as to who is to carry out the test. If no agreement is reached, the Agency shall specify who is to perform it. Again, this is subject to cost-sharing requirements.

[54] See Articles 27(5)–(7) and 30(3).

[55] Article 55 for the terminology of very high concern. This applies to substances on their own, in preparations and incorporated into articles.

[56] We will look later, in the next section on contestability, at how the content of this list is to be determined. At the time of adoption of the regulation, the Annex was empty.

[57] Article 55 and Article 62(4)(e).

[58] See Article 62 specifying the information to be submitted, and especially Article 62(4)(d). Note Article 62(5)(a) leaving applicants the discretion as to whether to submit a socio-economic analysis.

[59] Article 60(2).

show that estimated exposure falls within the relevant exposure limits, and that the likelihood and severity of an event occurring due to the physico-chemical properties of the substance is negligible.[60] The relevant authorities will assess whether the applicant has made his case, taking expert advice, but it is plain that the burden of proof falls upon the applicant.[61]

There is perhaps more ambiguity about burden of proof for those substances made subject to different authorization conditions. For PBTs/vPvBs, and substances for which no exposure limits can be determined, authorization will only be granted 'if it is shown that socio-economic benefits outweigh the risk to human health or the environment', and where there are no suitable, less risky alternatives.[62] The regulation is not explicit in identifying who bears the burden of demonstrating this. While an analysis of alternatives must be submitted as part of the application package, it is up to the applicant to decide whether to submit a socio-economic analysis.[63] While the various information submitted by the applicant will represent an element to be considered in reaching a decision, ultimately the decision will take into account all available information on risk and on the availability of alternatives.[64] A positive authorization decision may in principle be reached on the basis of information not brought forward by the applicant. Still though, in essence, the role of the applicant will be to persuade the regulator that the conditions for authorization are met, and to refute available information which might suggest otherwise.

Authorization, like registration, places an information burden on industry. In the case of authorization, it does so by reference to substantive standards, such as that represented by the concept of adequate control. Information and arguments submitted by industry in the course of authorization will not be taken at face value. It will be evaluated by experts including, where they deem it appropriate, in the light of additional information submitted by third parties.[65] Equally though, the claims of the experts will not always go unchallenged, with applicants enjoying an opportunity to make further comments along the way.[66] For the

---

[60] Section 6.4, Annex I. The exposure limits are in the form of Derived No-Effect Levels for human health (the level of exposure to the substance above which humans should not be exposed. See sections 1.0.2 and 1.4 of Annex I); and Predicted No-Effect Concentrations for environmental effects (see sections 3.0.1 and 3.3.1, Annex I).

[61] Note that Article 61(1) provides that the applicant submit an updated CSR as part of the process for the review of authorizations '[i]f he can demonstrate that the risk is adequately controlled'. 'Demonstrate' clearly places the burden of proof on the applicant. It is suggested that 'document' does so also.

[62] Article 60(4). As we will see below, the suitability of alternatives implies an assessment of their technical and economic feasibility for the applicant.

[63] Recall Article 62(5).

[64] Article 60(4).

[65] The procedure for authorization will be considered more fully below. By experts what is meant here are the relevant committees on Risk Assessment and Socio-Economic Analysis. See generally Article 64, especially Article 64(3) and (4).

[66] Article 64(5) ensuring that the EU administrative law requirement of right to be heard is complied with.

substances concerned, the default position is one of prohibition. It is not for the regulator to justify this, but for the potential user to challenge it. It is only when the potential user succeeds in making a prima facie case that the conditions laid down are met, that the regulator will incur a burden of justification to justify a ban.

3. Core Element: Contestability

One of the most intriguing, and potentially important, aspects of REACH is the way in which it combines harmonization with provisionality and contestability. The now dominant model of environmental regulation in the EU, premised upon substantive flexibility and proceduralization, is not well suited to product regulation in areas where uniformity is required in order to ensure the proper functioning of the European internal market. This presents a dilemma. Harmonization may be anticipated to come at the price of sacrificing the benefits of a more flexible approach. There is an upside to flexibility in that it is tolerant of regulatory diversity, and hence creates opportunities for regulatory learning on the basis of diverse Member State experience in implementation. Harmonization, on the contrary, by definition demands a singular response and would seem to exclude a role for Member States as laboratories for regulatory experimentation and learning. REACH, though, takes an imaginative step towards addressing this dilemma, and does so by inculcating the regime with the overlapping values of provisionality and contestability.

REACH takes Article 95 as its legal basis, and takes the form of exhaustive harmonization. Substances which comply with the regulation can be traded freely in the EU.[67] Recourse to the treaty-based free movement exceptions will be pre-empted.[68] But still, the Member States are not silenced. The regulation, and to a lesser extent the Treaty, preserve spaces for Member State contestation of the trade/regulation bargain struck. The contours of that bargain are provisional, and Member States are empowered in multiple ways to seek to destabilize it. There remains space for local knowledge to be integrated in a bid to provoke regulatory change. This local knowledge having been presented, the burden then falls on the EU decision-maker to justify its decision in relation to it, including from the point of the regulation's broad objectives, and in a manner which is compatible

---

[67] Article 128.
[68] Article 129. But note that, consistent with the case law of the European Court, recourse to these exceptions will be precluded only to the extent that the issue is covered by the harmonizing measure. This is apparent from Case 5/94 *Hedley Lomas* and reinforced by Article 128(2). So, for example, if a Member State wanted to restrict a chemical substance more strictly, for reasons not covered by REACH, it would remain free to do so, subject to the demands of other legislation, and with complying with the conditions in the Treaty-based free movement exceptions.

with the precautionary principle which underpins the regulation as a whole.[69] To this end, five particular points of entry are established in the regulation, and a further point of entry exists in the treaty. Each will be discussed in turn.

A. Substance Evaluation

It is for the European Chemicals Agency (the Agency) to draw up a list of priority substances for evaluation by Member States.[70] It will draw up a three-year rolling action plan of substances to be evaluated. Using risk-based criteria developed in cooperation with Member States,[71] and on the basis of an opinion from the Member State committee,[72] the Agency shall include substances in respect of which there are grounds for considering that they constitute a risk to human health or the environment. A Member State may notify the Agency at any time of a substance not on the list, whenever it is in possession of information that suggests that it is a priority for evaluation.[73] The Agency has the final word, and its decision is not subject to appeal before the Board of Appeal.[74] But still there is room for Member State contestation, placing a burden of justification on the Agency to explain why it has rejected the Member State's suggestion. The Agency *shall* include substances where there are grounds in any appropriate source of information, including presumably in a Member State petition, that it presents a risk, and hence a refusal to add a substance at the behest of a Member State would be subject to judicial review before the courts.

B. Substances Requiring Authorization

REACH establishes a prior authorization requirement for substances of 'very high concern'.[75] Unlike the registration requirement, this bites regardless of production volume. The regulation establishes definitively that this will apply to certain categories of substance. These include those which are category 1 or 2 carcinogens, mutagens, or reproductive toxicants.[76] It also applies to PBTs and vPvBs in so far as they meet the criteria set out in Annex XIII.[77]

Beyond this, however, the concept of a substance of very high concern remains open-ended. Additional substances may be subjected to this prior authorization requirement when they meet certain criteria and when they are added to the Annex XIV list.[78] The applicable threshold is a high one. It captures substances in

---

[69] See Article 1(1) for a statement of these objectives, and Article 1(3) for the inclusion of the precautionary principle.
[70] See Articles 44–48.
[71] These criteria shall consider hazard information, exposure information, and tonnage. See Article 44(1).
[72] See Article 44(2).
[73] Article 45(5).
[74] Article 93.
[75] Article 55.
[76] Article 57(a–c).
[77] Article 57(d–e).
[78] Article 57(f).

respect of which there is scientific evidence of probable serious effects to human health or the environment and which give rise to a level of concern which is equivalent to those of the other substances specifically listed.[79]

The process for adding or removing substances to the Annex XIV authorization list is a complex one. It culminates in the adoption of a decision on the basis of the regulatory committee with scrutiny procedure, the new comitology procedure enhancing the powers of the European Parliament.[80] Along the way Member States are empowered to push in the direction of the inclusion of additional substances, by arguing in favour of their inclusion on a preliminary candidate list.[81] In particular any Member State may choose to submit a dossier for substances which it believes meet the various criteria laid down, to forward this to the Agency. Interested parties will be given an opportunity to comment on this. In the absence of any such comments from interested parties, or from the Agency itself, the substance in question will be added to the candidate list. Where comments are received the matter will go to the Member State committee. In the absence of unanimity in this committee, a decision to add the substance to the candidate list of substances requiring authorization will be taken on the basis of a regulatory committee procedure, namely the older procedure not involving the European Parliament to any significant degree.[82] Of course, inclusion on the candidate list at the behest of a Member State does not imply inevitable inclusion on the final list. What is less clear is whether inclusion on the candidate list is a prerequisite for inclusion on the final list. This is an important question given the different procedures which apply: regulatory committee for the candidate list and regulatory committee with scrutiny for the final list, reflecting a greater role for the European Parliament in respect of the adoption of the final list. The capacity of the European Parliament to use its veto powers to bargain for the inclusion of additional substances will be greatly curtailed if the final list may only include those previously included on the candidate list, in respect of the drawing up of

---

[79] Substances with endocrine disrupting properties, or PBTs and vPvBs not meeting the Annex XIII criteria are given by way of examples of the kind of substances which might meet this threshold.

[80] Article 58(1) and Article 58(8). This is the procedure referred to in Article 133(4) of the Regulation. It refers to Article 5a of Council Decision 1999/468, which was recently introduced by Council Decision 2006/512. A full discussion of the various comitology forms is outwith the scope of this chapter. Suffice it to note that under this new procedure the European Parliament, like the Council, can veto the Commission's draft decision, on the basis that it exceeds the implementing powers conferred by the basic instrument (REACH in this case), or that the draft decision is not compatible with the aim or content of that basic instrument, or does not respect the principles of subsidiarity and proportionality. This is an important constitutional development as it puts the Parliament on a more even footing with Council in the adoption of implementing (executive) acts, as is already the case in respect of legislative enactments in most areas under the EC Treaty.

[81] See Article 59 which lays down the opportunities and obligations of Member States in this respect, and the relevant decision-making procedure.

[82] That is to say that the candidate list will be adopted in this event by regulatory committee procedure, whereas the final list will be adopted on the basis of a regulatory committee with scrutiny procedure. The regulatory committee procedure is that referred to in Article 133(3) and this, in turn, refers to Article 5 of the comitology decision, supra n. 80.

which the European Parliament is not much involved. Also, in that the Member States' contestatory powers bite only in relation to the candidate list, the more preliminary authority enjoyed by this, the more influential their contestation may be anticipated to be.

To my mind it seems unlikely that the final list must draw exclusively from substances included on the candidate list.[83] Were this to be the case, priority substances included in an Agency recommendation for inclusion in the final list[84] would have to be ignored if that substance did not also appear on the candidate list for inclusion. It is notable in thinking about this, that the Agency does not enjoy an autonomous right of initiative in relation to the candidate list. Only the Member States and the Commission may set in train the process of seeking to have a substance included on this candidate list.[85]

C. Restrictions

Annex XVII contains a list of restrictions concerning the manufacture, use or placing on the market of substances. This may be amended where there is an unacceptable risk to human health or the environment which needs to be addressed at Community level.[86] Member States may seek to contest the boundaries of Annex XVII by setting in train a procedure to consider its amendment, where it considers that the risk presented by a substance is not adequately controlled and needs to be addressed.[87] Where this demonstrates that action on a Community-wide basis is necessary, it shall submit this dossier to the Agency in order to initiate the restrictions process. The detail of the process for the adoption of new restrictions will be considered alongside the process for the adoption of authorization decisions below.

D. Harmonized Classification and Labelling

The regulation provides for harmonized classification and labelling of certain substances, and specifically for CMR (categories 1–3) and for respiratory sensitizers.[88] These shall 'normally' be added to Annex I of Directive 67/548. Harmonized classification and labelling may also be provided for 'other effects' on a

---

[83] Although it is clear that this could be implied by the language of Article 59(1) which presents the role of the candidate list as being such as to identify which substances meet the Article 57 criteria.
[84] See Article 58(3) and (4).
[85] Article 59(2) and (3). When the Commission exercises its right to initiate this procedure, it will ask the Agency to prepare a dossier on its behalf. But the Agency cannot, independent of instructions from the Commission, move to the preparation of a dossier of this kind.
[86] Article 68(1).
[87] Also the Commission may request the Agency to prepare a dossier of this kind. The Agency will do so when it considers that there is a risk to human health or the environment which is not adequately controlled. See Article 69(1) and (2).
[88] Article 115.

case-by-case basis if justification is provided demonstrating a need for Community-level action. The competent authorities of Member States may submit proposals to the Agency to this effect.[89] The Agency committee for risk assessment will adopt an opinion on this proposal, giving the parties concerned the opportunity to comment, and the Agency shall forward the opinion and comments for the adoption of a decision under Article 4(1) of Directive 67/548.[90]

E. Member State Safeguards

Article 129 establishes a Member State safeguard clause. This provides for the adoption of provisional (maximum 60 days) protective measures where a Member State has justifiable grounds for believing that urgent action is essential to protect human health or the environment. This is subject to an oversight procedure, whereby the Commission, acting on the basis of a regulatory committee procedure, will decide whether to authorize or approve the provisional measures. Where the provisional measure is approved, the Commission shall consider whether the regime as a whole needs to be adapted, and the Member State in question is required to initiate a Community-wide restrictions process as discussed above.

F. Article 95 EC[91]

Whenever an EU legal act is based on Article 95 of the EC Treaty, certain 'opt-outs' are constitutionally guaranteed by Article 95(4) and 95(5). Though subject to a common procedure, the two sub-paragraphs differ in their scope. Article 95(5) relates to the adoption by Member States of a new regulatory measure after the enactment at EU level of the harmonization measure in question. This is narrowly drawn, justifiable only on grounds relating to the protection of the environment or working environment, and even then only due to the emergence of new scientific evidence relating to a problem specific to that Member State. The Article 94(4) opt-out, by contrast, relates to the maintenance of national measures in place even before the enactment of the EU harmonizing act. The range of grounds on which Member States may rely is broader, including those specified in Article 30, as well as environment and working environment. Thus, for example, this would include public health.

---

[89] As with the candidate list for authorization and new restrictions proposals, the Member State shall submit an Annex XV dossier setting out the justification.

[90] Article 115(2). The formulation here in terms of participation is different, concerned as it is with 'parties concerned' rather than 'interested parties'.

[91] Note also Article 67(3) providing that Member States may maintain any existing and more stringent restrictions in force, relative to Annex XVII, until 1 June 2013, subject to their being notified according to the Treaty. The deadline for notification is not set out but it must be before 1 June 2009, as the Commission is required to publish an inventory of notified restrictions by this date. Many thanks to Peter Oliver for clarifying this point for me.

Under the Article 95 opt-outs, it is for the Commission to approve or reject the national provisions in question, usually within a six-month period. For this reason, Member States are obliged to notify their measures to the Commission. Any measure not so notified will be deprived of legal effects. In the course of reaching its decision, the Commission will scrutinize compliance with the various conditions, and verify also that the measure does not constitute a disproportionate obstacle to the functioning of the internal market.[92] As the recent Austrian case concerning genetically modified organisms shows, the Commission's decision will be susceptible to judicial review.[93] Where a Member State measure is authorized, it will be for the Commission to examine whether to propose an adaptation to the harmonizing measure in relation to which the Member State is opting out.[94]

The Article 95 opt-outs (or their predecessors) have been invoked nineteen times by Member States in relation to chemicals, with Member State measures receiving approval on thirteen occasions.[95]

Article 95 has something in common with the other channels for contestation in REACH; not least the fact that a successful Member State opt-out acts as a catalyst for the Commission to consider proposing an adaptation to the regulatory bargain for the Community as a whole. By contrast to the other mechanisms though, it is the Commission alone which is responsible for making a determination about whether to reject or approve the application in question. In this setting the Commission is unconstrained by either comitology or other consultation requirements.[96]

G. Contestability: The Limitations

It is apparent from the above that the Member States enjoy multiple opportunities to contest the harmonized settlement represented by the regulation. They may do so by exploiting the opt-out possibilities established under Article 95 EC or by the safeguard clause in REACH. These possibilities are quite narrowly drawn. More particularly, and more unusually, Member States may also contest the boundaries of the regulatory settlement by setting in train a process of

---

[92] To preserve the useful effect of these opt-outs, the Commission read the 'no obstacle' requirement as meaning 'no disproportionate obstacle'.
[93] Joined Cases C-439/05 P and C-454/05 P *Land Oberösterreich and Austria v. Commission* (judgment of 13 September 2007).
[94] So, for example, in relation to the Dutch example above, the Commission had proposed an amendment to the EU framework governing creosote and creosote-treated wood.
[95] See the list at: <http://ec.europa.eu/enterprise/chemicals/legislation/derogations/index_en.htm>.
[96] For a recent discussion of the limits to even the applicant Member State's right to be heard, see Cases C-439/05 P and C-454/05 P *Land Oberösterreich and Republic of Austria v. Commission*. Here the ECJ places emphasis upon the importance of achieving a speedy resolution of such issues, without time-consuming information exchange (para. 41).

reflection, according to which there will be detailed and reasoned contemplation of the need for change. Member States are accountable in making recourse to these procedures, not least through the frequent requirement that their bid for change be accompanied by a dossier which lays down the reasons for this and the evidence on which the conclusion of that Member State is based. The contours of such dossiers are set out in Annex XV.

Like the Member States, the Commission too enjoys certain opportunities to launch such processes. This is most apparent when it comes to the drawing up of the candidate list of substances requiring authorization, and in relation to the restrictions process.[97] That said, the Commission's powers are more circumscribed, notably in respect of the restrictions procedure.[98] Here, the conclusion of the Commission that a substance poses a risk will be examined by the Agency. The Agency will not automatically draw up a dossier upon request by the Commission but will do so only where it shares the Commission's view that the substance poses a risk which is not adequately controlled, and that action on a Community-wide basis is necessary.[99]

Still, this leaves the question of which other actors enjoy these formal opportunities to destabilize the settlement, and to contest the boundaries of the regulation. Two observations are pertinent here.

First, the Agency's autonomous capacity to do so is limited. While it is a central actor in all of the decision-making processes, regardless of where the initiative to contest originates, its powers to contest directly the regime's existing boundaries are constrained. Of course it enjoys considerable autonomy in drawing up the rolling action plan of substances to be made subject to evaluation.[100] And it is empowered to recommend substances for inclusion on the final list of those requiring authorization.[101] Beyond this it is an agent of the Commission, being made responsible for the preparation of a dossier, at the Commission's request, in respect of the candidate list of substances and the restrictions process. It is particularly significant that the Agency enjoys no independent right of initiative in seeking to launch a restrictions process.

Second, although interested parties are offered multiple opportunities to make their views known in the course of adopting implementation decisions, including

---

[97] See Article 59(2) and Article 69(1).

[98] The Commission does not enjoy the same powers to contest in respect of substance evaluation, harmonized classification and labelling, or the safeguard clause. Though recall the discussion above about the emergency powers of the Commission in the context of review of authorization decisions.

[99] Article 69(1) and (2). This is in contrast to the Commission's contestatory role in respect of authorization, where a Commission request to the Agency to draw up a dossier to consider a substance for inclusion on the candidate list will not be second-guessed by the Agency but carried out without question. See Article 59(2). This probably reflects the Agency's own role in recommending substances for inclusion on the final list under Article 58(3). This provides the Agency with the opportunity to make its own views known.

[100] See again Articles 44–48.   [101] Article 58(3).

in the course of drawing up the candidate and final lists of substances requiring authorization, and in relation to the restrictions process, such interested parties are not empowered themselves to set in train the various reflection processes. Neither industry actors or NGO voices, nor academics or other experts, are given formal recognition as catalysts for revision in the light of new information or better understanding. As noted, the boundaries of the regime are porous. But the range of actors which can seek to alter them is noticeably confined. It is to be regretted that other persons or organizations do not enjoy the capacity to bring forth evidence-based dossiers to launch formal inquiries about the need for change. It would be reasonable to think that the Agency might serve as an intermediary in this event, to examine dossiers submitted, in order to ensure that frivolous claims do not progress. In much the same way as the Agency will examine the credibility of the Commission's claims in seeking to launch an inquiry into the need for new restrictions, so too it could serve as a gatekeeper for attempts at de-stabilization brought forth by civil society actors.

H. A Note on Authorization and Restriction Procedures and Criteria

It has been illustrated above that Member States enjoy multiple opportunities to contest the boundaries of the regulation. They enjoy very limited authority to impose stricter standards than those laid down in REACH, but they may in various ways set in train a procedure to consider ratcheting those standards up. Ultimately, decisions will be taken in accordance with the procedures laid down. Important differences exist between these procedures. This may be most starkly illustrated by reference to the distinct procedures for the adoption of authorization and restrictions decisions.

1. *Authorization*

(a) **Procedure**

The Agency committees shall issue draft opinions on an application for authorization.[102] The Committee for Socio-Economic Analysis shall take into account any information on alternative substances or technologies submitted by interested third parties.[103] Interested parties are not invited to comment more generally. However, the applicant is given an opportunity to comment, and the

---

[102] See Article 85 on the Committee for Risk Assessment and the Committee for Socio-Economic Analysis. These comprise members appointed by the Agency's Management Board from a list of nominees established by the Executive Director on the basis of Member State nominations. At least one nominee put forward by each Member State will be appointed, but not more than two. Members will be appointed for their role and experience in performing the Agency's task. The committees shall aim to have a broad range of relevant expertise, and to this end each committee may co-opt up to five additional members, chosen on the basis of their specific competence.
[103] Article 64(4)(b) and 64(2).

committees shall consider the comments, taking this argumentation into account when appropriate. It shall submit its opinions, with the written argumentation attached, to the Commission, the Member States, and the applicants.[104] The Commission shall adopt a draft decision and adopt its final decision on the basis of a regulatory committee procedure.[105]

(b) Criteria

The Article 60, para. 2 criteria apply to authorizations for Annex XIV substances except where these are PBTs or vPvBs, or where it is not possible to establish a DNEC or a PNEC. In this case, authorization will be granted if it is established that the risk to human health and the environment is adequately controlled in accordance with section 6.4 of Annex I.

For other authorization substances, Article 60, para. 4 will apply. For these substances, authorization will only be granted if it is shown that socio-economic benefits outweigh the risk to human health or the environment, and if there are no suitable alternative substances or technologies. In considering the availability of alternatives account will be taken of the overall risk profile of an alternative, and of its technical and economic feasibility for the applicants.[106]

2. Restrictions

(a) Procedure

When a Member State or the Agency prepares a restrictions dossier demonstrating the need for a Community-wide restriction, interested parties will be given an opportunity to submit comments and/or a socio-economic analysis.[107] The Agency Committee on Risk Assessment shall draw up an opinion taking into account these comments (but not the socio-economic analysis).[108] The Agency Committee for Socio-Economic Analysis shall formulate a draft opinion, taking into account any socio-economic analyses submitted by interested parties. It will invite interested parties to comment upon this. It will adopt its final opinion taking into account earlier socio-economic analyses submitted, and subsequent comments received.[109] These committee opinions will be submitted to the Commission.[110] If the conditions for imposing a restriction are satisfied (see below) the Commission shall prepare a draft amendment to Annex XVII, taking its final decision in accordance with a regulatory committee with scrutiny procedure. Where its draft amendment diverges from the original proposal (put forward by the Member State or Agency), or does not take the opinions of the

---

[104] Article 64(5).   [105] Article 64(8).
[106] Article 60(5)(b).   [107] Article 69(6).
[108] Article 70.   [109] Article 71(1) and (2).
[110] Article 72(1). In the absence of an opinion from either or both of the committees the Agency shall inform the Commission and state reasons for the absence.

committees into account, the Commission shall annex a detailed explanation of the reasons for the differences.[111]

(b) Criteria

Restrictions will be imposed when there is an unacceptable risk to human health or the environment, arising form the manufacture, use, or placing on the market of substances, which needs to be addressed on a Community-wide basis.[112]

3. Comparison of the authorization and restrictions procedure

There are important differences in the procedures for the adoption of authorization and restriction decisions.

First, interested parties enjoy a broader right to submit comments under the restrictions procedure. Under the authorization procedure their right to intervene is restricted to the submission of information about alternative substances or technologies.

Second, authorization decisions are adopted by a regulatory committee procedure, whereas restriction decisions are adopted by regulatory committee with scrutiny. Thus the European Parliament is greatly empowered in the latter.

Third, an individual applicant enjoys a right to be heard under the authorization procedure.

Fourth, under the restrictions procedure the Commission is not obliged to take the Agency committee opinions into account. In this event it must offer a reasoned justification. While there is neither an obligation to take the committee opinions into account under the authorization procedure, nor any obligation to offer reasoned justification for departure, the case law of the European Court would seem to imply that both are necessary in any case. In each case the case law would suggest that departure from these expert opinions is circumscribed by an obligation to offer reasons and evidence for so doing. Where the committee offers scientific advice, departure is countenanced only where the Commission can offer scientific justification at a level which is commensurate with the committee opinion.

Taking the first two differences together, it is clear that there are opportunities for public participation, directly and indirectly through democratically elected representatives, enhanced in relation to the restrictions procedure, by comparison to the authorization procedure. Yet, particularly for the most risky substances requiring authorization under Article 60(4), the conditions for authorization are anything but purely scientific. Wide judgement is called for, in balancing costs and benefits, and in assessing the suitability of putative alternatives, including their economic feasibility for applicants.

[111] Article 71.
[112] Article 68(1) and Article 73.

Elsewhere, interested parties are given greater opportunities to comment, when the power of the European Parliament in comitology declines. Thus, in drawing up the candidate list for authorization, interested parties are allowed to comment, but the power of the European Parliament is diminished by recourse to the traditional regulatory committee procedure.[113] In drawing up the final list, by contrast, the European Parliament is involved by dint of the scrutiny procedure, but the individual right to comment is curtailed.[114] Both the restrictions and authorization procedures are unusual; the former by combining parliamentary involvement with direct public participation, and the latter by curtailing both.

4. Core Element: Substitution

One consequence of REACH is that it will narrow, and ultimately close, the regulatory gap between new and existing chemicals. In the past the additional regulatory burden attaching to new chemicals has created a perverse incentive, favouring the continued use of existing chemicals, even where these existing chemicals are risky, and even where safer alternatives exist or could readily be developed. REACH seeks to alter this incentive structure. It does so most obviously by requiring the phased registration of existing chemicals. And it does so expressly when it comes to the authorization and restriction of risky chemicals.

As noted in the previous section, certain very risky substances will require prior authorization. Those applying for authorization are obliged to analyse the availability of alternatives and to consider their risks, as well as the technical and economic feasibility of substitution.[115] All applications for authorization are to include this analysis. This procedural obligation is combined with a substantive requirement that, for many of the substances requiring authorization, approval will only be granted if there is no suitable alternative substance or technology.[116] Suitability will be determined having regard to overall risks, as well as to the technical and economic feasibility of alternatives for the applicants.[117] It will be determined within the framework of the authorization procedure set out earlier in this paper. It will be for the Agency Committee for Socio-Economic Analysis to assess alternatives. It will do so having provided an opportunity for interested

---

[113] See Article 59.
[114] Article 58(1). Individuals do have an opportunity to comment on the Agency recommendation identifying priority substances. See Article 58(4).
[115] Article 55.
[116] Article 60(4). This requirement applies when para. 4 rather than para. 2 of Article 60 is applicable. This is the case for PBTs and vPvBs, and for other high-risk chemicals included on the final authorization list where it is not possible to determine a level of exposure threshold above which humans should not be exposed, or an environmental concentration threshold below which adverse environmental effects are not expected to occur. See Annex I, 1.0.1. and 3.0.1.
[117] Article 60(5).

third parties to submit information on alternatives, and having given the applicant a further right to submit comments following the publication of a draft opinion.[118] While the Commission is not bound by the Agency opinion, it is obliged to justify departure from it, including in relation to the suitability of alternatives.[119]

There is something here which is resonant of Wendy Wagner's proposal for 'competition-based regulation' in relation to chemicals.[120] According to this, the role of a regulator should be to adjudicate upon claims of environmental superiority, by inviting competitor companies to petition for certification of their product as superior, and for the restriction or banning of inferior products.[121] Superiority determinations should take shape within the framework of an adversarial procedure which is robust in its ability to guard against over-blown claims or misstatements of fact.

It might be argued that REACH creates just such an inclusive forum for adjudication on the suitability of alternatives, and for the banning of substances where a 'superior' alternative is shown to exist. The process of adjudication is mildly adversarial. The Agency does not take the word of the applicant as gospel, but invites the submission of rival information on alternatives from 'interested parties', including on the part of competitors. In certain respects, REACH goes further than Wagner proposes. For example, even more expensive alternatives may be deemed suitable, so long as they remain within the realm of economic feasibility. In other respects, however, REACH falls short of the Wagner paradigm. It applies only to high-risk substances, and not even to all of them.[122] Also, while, as noted in the previous section, there is some fluidity in what is to count as a high-risk substance, the initiative to expand the list rests in the hands of the Commission and Member States (and to a lesser extent the Agency). Competitors are not empowered to petition for a substance to be included on this list, and hence are not able to play a role in galvanizing the superiority adjudication.

This emphasis upon substitution is further reflected in relation to the restrictions process. Restrictions may be imposed where a substance presents an unacceptable risk to human health or the environment. Consistent with Wagner's suggestion in relation to the US Toxic Substances Control Act (TCSA),[123] here what counts as an unacceptable (unreasonable under TCSA) risk will be determined taking into account the socio-economic impact of a

---

[118] Articles 64(2) and (5).
[119] This obligation is inherent in the administrative law of the EU. See Case T-11/99 *Pfizer*.
[120] Supra n. 8.
[121] What counts as superior is stated at p. 22: 'If a company establishes that their product is significantly safer to public health or the environment than a competitor product for an identified set of uses, and is available at roughly the same price per application, then the product could be certified as competitively superior for those uses.'
[122] Recall the discussion above about the different conditions for authorization under Article 60(2) as compared to Article 60(4).
[123] Wagner, supra n. 8, p. 20.

restriction, including in the light of the availability of alternatives.[124] Once again, it is for the Agency Committee on Socio-Economic Analysis to draft an opinion on the proposed restriction. In so doing, it is obliged to take into account the analyses or information submitted by interested parties, including presumably information on the availability of alternatives.[125] While once again it is not within the gift of a competitor to set a restrictions process in motion, this process having been launched by a Member State or by the European Commission, a competitor is able to play an important role in providing information and in contesting received wisdom, including in relation to the socio-economic impact of putative restrictions, in the light of their knowledge of available alternatives.

The capacity of REACH to re-calibrate incentive structures in chemicals regulation, in favour of the development of 'green(er)' alternatives, will depend in large part on the way in which the substitution dynamic in authorization and restrictions plays out in practice. It will depend upon the willingness of the various political actors to use the vague language (suitable or available alternatives) to sanction the existence of workable alternatives, and upon the willingness of independent scientists and particularly competitor companies to come forward to play a role in the quasi-adjudicative process for certifying the existence of alternatives. REACH asks much less of competitor companies than Wagner would like. It merely asks them to engage in a process the impetus for which comes from elsewhere. It does not ask them, or indeed allow them, to instigate this process themselves. For a competition-based approach to regulation to work, competitors must be willing to succumb to a strategy of 'divide and rule' on the part of the regulator, and to be willing to attest formally to the superiority of their product relative to that of an industry rival. In an admittedly attenuated form, REACH may provide a test case for this. It will provide an occasion to assess the willingness of industry to participate in authorization or restriction processes concerning a competitor's product, and the willingness of industry to adduce evidence about alternatives in a bid to provoke a negative authorization decision or the imposition of restrictions in relation to that competitor's product. Time will tell.

5. Core Element: Provisionality

A. Review of Authorizations

As will be discussed in more detail below, substances of very high concern will be subject to an authorization requirement, even at very low volume.[126] Where

---

[124] Article 68(1). [125] Article 69(6)(b) and (71)(1) and (2).
[126] See Section B above.

authorizations are granted, these will be subject to a time limited review, and will normally be subject to conditions, including monitoring.[127] They will also be subject to the all-important proviso that the authorization holder will incur an absolute obligation to ensure that exposure is reduced to as low a level as is technically and practically possible.[128] This is somewhat reminiscent of the IPPC concept of best available techniques, a standard which gives rise to an evolving rather than an absolute obligation.[129] What is interesting here is that while socio-economic considerations will be directly relevant in the authorization process, and while the availability of substitutes will be subject to an economic feasibility test, this obligation to attain lowest exposure is not subject to any economic or cost-benefit balancing test. Admittedly it is only concerned with one element of the overall risk package—exposure as opposed to hazard—but even so it provides a strong instrument to the regulator in performing its review functions.

Authorization decisions will remain valid until they are amended or withdrawn subject to the obligation on the authorization holder to submit the review report discussed above, at least eighteen months before the expiry of the authorization time limit.[130] Review may be routine, associated with the time limit, or ad hoc. The Commission may conduct an ad hoc review at any time, if the circumstances of the original authorization have changed in such a way as to affect risk to humans or the environment, or if new information on possible substitutes becomes available.[131] Authorization may be withdrawn or amended, following a routine or ad hoc review, if under the changed circumstances it would not have been granted, or if suitable alternatives have since become available.[132] Where there is a serious and immediate risk for human health or the environment, authorization may be suspended pending the review.[133]

It is significant that review of authorization provides an opportunity to link REACH to other aspects of the EU's environmental programme. It is explicit in providing for the possibility of review in so far as IPPC environmental quality standards are not met or environmental objectives under the Water Framework Directive are not achieved.[134] In this way, REACH is somewhat parasitic for the attainment of its objectives, upon the monitoring regimes established by other instruments. The Water Framework Directive, in particular, lays down elaborate

---

[127] Article 60(8). The authorization will specify the time-limited review period and any monitoring arrangements.
[128] Article 60(10).
[129] Directive 96/61, Article 2(12).
[130] Article 61(1).
[131] Article 61(2).
[132] Article 61(3).
[133] Article 61(3). This is very much the Commission's equivalent to the Member State safeguard clause in Article 129 to be discussed below.
[134] Article 61(4) and 61(5). See also Article 61(6) which provides for the automatic withdrawal of authorization where the substance is subsequently prohibited or restricted under Regulation 850/2004 on persistent organic pollutants (POPs).

and detailed requirements on the monitoring of water status, and on the regular reporting by Member States of the results of this.[135]

B. Reporting/Review/Revision

Along the way, in the lengthy course of the adoption of REACH, difficult and contested choices had to be made. It is striking, in the light of this, to note that the regulation is self-consciously provisional, in that it explicitly contemplates the circumstances for its own revision. It does so not merely in the abstract, or in general terms, but on the basis of carefully defined questions for inquiry, and in an institutional setting which confers specific review responsibilities on the Commission.[136] During the twelve-year period following the entry into force of the regulation, the Commission is charged at different points with reviewing specified elements of the regime, and with bringing forward proposals for reform. In keeping with Article 131, amendments to the regulation's multiple and extensive annexes may be achieved by the Commission acting on the basis of a regulatory committee with scrutiny procedure. For amendments to the main body of the text, the Commission will be compelled to bring forward proposals for the adoption of legislative acts.

The range of questions to be addressed by the Commission is extensive, with nine specific tasks set out. The Commission is required to review the scope of certain obligations, including some which were among the most controversial during the enactment phase. Thus, for example, the Commission is to consider whether in the future to require a chemical safety report for substances produced at relatively low volume, below the current 10 tonne threshold;[137] or whether to modify the Annex XIII criteria for identifying PBTs and vPvBs.[138] The Commission will consider whether to extend the more rigorous authorization procedure to endocrine disrupting substances;[139] and whether to extend the duty to

---

[135] Directive 2000, Annex V. This might go some way to addressing concerns raised that REACH is not sufficiently committed to environmental monitoring as part of its approach to identifying and regulating against risk. See McEldowney, S., 'EU Chemicals Regulation: A Foundation for Environmental Protection or a Missed Opportunity?' Vol. 4 *Yearbook of European Environmental Law* (OUP, 2005).

[136] See Article 138.

[137] Article 138(1) to be carried out by 1 June 2019, except in so far as it relates to CMR (categories 1 and 2) substances in which case it is to be completed more quickly, by 1 June 2014. Looking at the precise wording, this would seem to include a review of this requirement for substances not currently subject to registration because they are manufactured or imported in quantities of less than one tonne.

[138] This is to be carried out more expeditiously, within one year of entry into force, by 1 June 2008.

[139] This is to be carried out by 1 June 2013. It will be recalled that Article 60 establishes two procedures for authorization under para. 2 and para. 4. At present Article 57(f) substances are subject to para. 2 except in the circumstances laid down in Article 60(3). As such they may be authorized so long as the risks are adequately controlled, and without being subject to an alternatives assessment, or a balancing of socio-economic benefits against risk.

communicate information on substances in articles to dangerous substances not currently included on the prior authorization list.[140]

In all but one case, the Commission's review obligations are one-off, to be carried out by the date specified. In the odd case out, there would appear to be a regular, recurring, obligation on the Commission to review the information requirements for substances produced at a volume of 1–10 tonnes.[141] This review obligation is explicitly linked to the Commission's obligation to produce a five-yearly general report on the operation of the regulation.[142] This general report will be issued following the submission of implementation reports by both Member States and the Agency.[143]

This regular reporting cycle also provides an important opportunity for reflection and review. Member State reporting is an endemic feature of EU environmental law. In practice, the Commission shapes the submission of information by Member States by issuing a questionnaire.[144] This serves to facilitate transparent comparison of Member State approaches, and of their achievements to date by the Commission in its synthesis report. It is early days with REACH, and as yet no such reports have been issued. There is, however, experience in other areas which is illustrative of the potential inherent in reporting and review mechanisms of this kind. Over the years, the scope of the Environmental Impact Assessment Directive has been extended, and its obligations sharpened. Many of the most important amendments introduced have their origin in lessons learned on the basis of Member State and Commission reports.[145] Member States come to constitute laboratories for experimentation in the implementation of often open-ended obligations. Experiences are then

---

[140] Article 138(8). This is to be undertaken by 1 June 2019, taking into account practical experience in implementing the article.

[141] Article 138(3).

[142] Article 117(4).

[143] Article 117(1) and 117(2). Under Article 117(3) the Agency also incurs a three-yearly obligation to submit a report on the status of implementation and use of non-animal test methods and testing strategies. This information will assist the Commission in performing its review function under Article 138(3) discussed above. The existence of a reporting obligation of this kind acquired a legal significance in the recent *Mangold* case (Case C-144/04). This was a factor cited by the Court in justifying its decision to conclude that the German law in question should be set aside as contrary to an EU directive, even though the deadline for implementation has not passed. In so doing, it relied upon the *Inter-Environnement Wallonie* doctrine (Case C-129/96), concluding that the regular reporting obligation incurred by states invoking a time-limited extension, implied the existence of a standstill obligation, whereby Member States were precluded from adopting measures which would make it impossible to achieve compliance at the end of the time-limited period.

[144] The Agency Forum (see n. 146 below) is responsible for establishing an electronic information exchange procedure. As with the Water Framework Directive, it is likely that this will result in informally agreed templates for the submission of information in reports, which will make it easier for the Commission, and others, to spot gaps, and compare performance across Member States. On the Water Information System for Europe (WISE) see, <http://www.wise-rtd.info/>.

[145] For a fuller discussion see Scott, J. and Holder, J. 'Law and New Environmental Governance in the EU' in de Búrca, G. and Scott, J. (eds.), *Law and New Governance in the EU and the US* (Hart Publishing, 2006).

pooled, allowing examples of good and bad practice to inform the revision of the overall regime. The reporting and review cycle provides an institutional framework for the integration of on-the-ground experiences in implementation, and for the organized and transparent participation of Member States in shaping the revised framework binding on all Member States.

This remains important in the context of REACH, although it is in the form of a Regulation, and although, given its internal market focus, it promotes greater upfront harmonization across Member States than is frequently the case. Still, Member States enjoy sometimes substantial flexibility in implementation. This is notably the case when it comes to enforcement. Article 125 simply requires Member States to 'maintain a system of official controls and other activities as appropriate to the circumstances'. Article 126, in similarly enigmatic vein, provides that they 'shall lay down the provisions on penalties applicable for infringement...and shall take all measures necessary to ensure that they are implemented. The penalties provided for must be effective, proportionate and dissuasive.' It is then hardly surprising that it is stipulated that Member State reports shall include sections on evaluation and enforcement, including the results of official inspections, the monitoring carried out, the penalties provided for, and other measures taken during the reporting period.[146] In this way, the reports serve an accountability as well as a learning function, rendering transparent and comparable Member State performance in areas characterized by open-ended norms and flexibility in implementation.[147]

6. Core Element: Transparency

A. Information in the Supply Chain

A key premise of REACH is that effective risk management requires effective communication across the entire supply chain for chemicals. Communication should be two-way: down-stream to ensure the provision of up-to-date information necessary to facilitate safe use; and up-stream to allow for new,

---

[146] See also Article 127. This also states that the common issues to be covered in the reports shall be agreed by the forum. This is to be established by Article 86, and will be a successor to CLEEN, a chemicals enforcement network which has been in effective operation for over ten years. It shall comprise individuals selected by each Member State, an Agency representative, as well as representatives of the Commission, with stakeholders being invited to attend as observers. The forum may also co-opt up to a maximum of five additional members in a bid to ensure a broad range of relevant expertise among members.

[147] The activities of the Forum are laid down in Article 77(4) and include proposing, coordinating, and evaluating harmonized enforcement projects and joint inspections, identifying enforcement practices in enforcement, developing working methods and tools of use to local inspectors, developing an electronic information exchange procedure, and examining proposals for restrictions with a view to advising on enforceability.

on-the-ground, experience based information on risk and risk management to be fed back to the manufacturer or importer.[148]

At the core of the REACH system is the concept of the Safety Data Sheet (SDS). The requirements of this are set out in the Annex II 'guide', in a bid to ensure consistency and accuracy in the presentation of data. It is to include information on the intrinsic properties of substances; known uses; handling, storage, and disposal information; first aid, fire-fighting and accidental release measures; and exposure limits and control measures. These are designed 'to enable users to take the necessary measures relating to protection of human health and safety at the workplace, and protection of the environment'.[149]

The provision of an SDS is mandatory in only a limited range of situations. These concern the supply of substances or preparations which are dangerous, substances which are PBTs/vPvBs (meeting Annex XIII criteria), and substances included on the Article 59(1) candidate list for authorization.[150] Recipients may also receive, upon request, an SDS for certain preparations which contain substances which though not dangerous, pose human health and environmental hazards or which are PBTs/vPvBs not meeting the Annex XIII criteria.[151]

Outside of these situations, suppliers of substances on their own or in preparations still incur a duty to communicate specified information down the supply chain. This includes information about restrictions and authorization, but also more generally available and relevant information about the substance that is necessary to enable appropriate risk management measures to be identified and applied.[152]

The above is concerned with disclosure of information on substances on their own or in preparations. For substances in articles, controversially, different requirements apply.[153] Here, there will never be an obligation to prepare an SDS. For the most hazardous substances, identified on the basis of their inclusion on the Article 59(1) candidate list for authorization, there is instead merely a general, and relatively weak, disclosure requirement. This takes the form of a standard rather than a rule. It requires the disclosure of sufficient information to allow safe use. But it is subject to a proviso. Subject to an absolute obligation to provide information on the name of the substance, only information *available* to the supplier need be supplied.[154] The concept of availability is not defined.

---

[148] The discussion will focus here on downstream provision of information, but see Article 34 on the duty to communicate up-stream.

[149] Annex II, Recital 2. Where the relevant actor has been required to prepare a CSR as part of the registration or authorization process, it shall include the relevant exposure scenarios in an annex to the SDS. See Article 31(7). These only form part of a CSR for dangerous or PBT/vPvB substances.

[150] Article 31(1). The reference in Article 31(1)(c) to points (a) and (b) in the context of Article 59 (1) is quite confusing. This refers back to (a) and (b) of Article 31(1) and not Article 59(1) and reflects the fact that some substances included on the candidate list will also be dangerous and/or PBTs etc.

[151] Article 31(3). This is subject to specified minimum concentration thresholds.
[152] Article 32(1).   [153] Article 33.
[154] The concept of availability applies also under Article 32 concerning substances and preparations not requiring an SDS. But here the bottom-line disclosure requirement is more expansive, including information on authorization and restrictions.

More generally in thinking about the scope of obligations to supply SDS and other supply chain information requirements, it is important to contemplate the place of the final consumer. In thinking about this, it is crucial to be aware that information in the supply chain, including SDS, is to be supplied to recipients of substances on their own or in preparations. The definition of a recipient is given in Article 3(34) and is narrower than might otherwise be assumed. This means a down-stream user or distributor being supplied with a substance or a preparation. While it is plain that a consumer is not a distributor,[155] it transpires that neither does a consumer constitute a down-stream user for the purpose of the regulation. Only industrial or professional users are constituted as such.[156] Hence, when it comes to that part of the REACH package which is concerned with information in the supply chain (Title IV), consumer entitlement to receive information is somewhat lacking. Consumers are entitled to request such information as is necessary to allow for the safe use of an article containing a substance of very high concern, apart from information on the name of that substance, though this obligation is subject to the relevant information being 'available' to the supplier.[157]

It is then apparent that there are shortcomings or gaps in arrangements for the communication of information down the supply chain, and particularly in respect of consumers. These shortcomings are particularly pronounced in respect of substances in articles, where an SDS is never required. In view of these shortcomings, it is important to ask to what extent they are mitigated by other aspects of the REACH transparency package. To determine this, it is necessary to look at the rules laid down on harmonized classification and labelling on the one hand, and on public access to information on the other.

B. Classification and Labelling of Substances

By contrast to information in the supply chain which is directed in the main at professional and industrial users, harmonization of classification and labelling of chemicals aims primarily at the protection of consumers and workers.[158] This is true both for the current EU system and for the Globally Harmonized System of

---

[155] See Article 3(14). A consumer does not store a substance and place it on the market for third parties.
[156] Article 3(13).
[157] See Article 33(2). This is also subject to a concentration requirement. See also Article 31(4) which may create an incentive to provide consumers with information. Where the ultimate user is a consumer and not a professional or industrial user, provision of this information to the consumer will mean that an SDS will only have to be supplied upon request. This is important as consumers will be able to receive, upon request, information on the presence of substances of very high concern in articles.
[158] But note Article 35 which provides that workers and their representatives shall be granted access by their employer to SDSs and other information on substances and preparations that they use or may be exposed to in the course of their work.

Classification and Labelling of Chemicals (GHS) which the EU is in the course of implementing.[159]

The core function of both the old and new system is to harmonize hazard classes for dangerous substances and preparations (so-called mixtures under GHS), to draw up a list of substances with harmonized classifications for specific hazard classes, and to regulate packaging,[160] and the provision of information to consumers and workers through labelling,[161] including hazard statements,[162] precautionary statements, and pictograms.[163] There are also rules on advertising, to avoid misleading customers. Where a substance or preparation is not subject to harmonized classification, it must be classified in accordance with established hazard classes by the supplier prior to it being placed on the market. The regime also provides for a publicly accessible classification and labelling inventory, containing the relevant information submitted to the Agency as part of the registration or notification process.[164]

A detailed evaluation of GHS is outside the scope of this paper. What is crucial for our purposes is that it provides a framework for the protection of consumers, in a way that the supply-chain information requirements do not. But equally it is plain that the scope of application of this framework is limited in that it applies only to substances on their own or in preparations (mixtures). With one exception, it does not apply to substances in articles, even where these are intended to be released or where it is reasonably foreseeable that they might be in the course of normal use.[165] When it is recalled that SDS are never required for substances in articles, and that consumers may only receive available information on request, including at a minimum the name of the substance, there does still appear to be a significant gap in the protective regime constituted by the

---

[159] To this end, the provisions of Title XI of REACH on classification and labelling will be moved to the new classification and labelling regulation. See COM(2007) 355 final for the proposal for a regulation on classification, labelling, and packaging of substances and mixtures, which will implement in the EU the international criteria agreed by the UN Economic and Social Council.

[160] See, for example, Annex II, Part III which provides for child-proof fastenings and tactile warnings.

[161] See Annexes II–V of the proposed regulation setting out the information to be included. Note that this includes in Annex II supplemental labelling provisions derived from the existing EU regime and not yet provided for by GHS. Similarly, in Annex III, additional hazard statements are required for hazards not currently part of GHS. Annex VI runs to more than 900 pages, comprising the list of substances with harmonized classifications.

[162] See Annex III. These provide information on the hazardous properties of substances. For example 'Explosives, Unstable Explosives' or 'Flammable Gases' or, more extreme, 'Reproductive Toxicity', 'Acute Toxicity', or 'Carcinogencity, Hazard Category 1A, 1B'.

[163] See Annex V, reproducing the GHS hazard pictograms, such as flames, skull and crossbones, exclamation marks, a highlighted human chest, or fish in the environment.

[164] See Article 114 REACH, and Article 113 specifying the information to be included. At present the notification obligation under REACH applies both to the registered substances and preparations and to dangerous substances even when these are not subject to a registration obligation. Under the new proposal, registered substances will be treated as having been notified as the registration process already requires the submission of information on classification and labelling.

[165] The exception concerns flammable aerosols.

transparency requirements.¹⁶⁶ How significant depends in part upon how the ambiguous line separating articles from preparations will be drawn. Assistance should be provided in this regard by the REACH guidance document on substances in articles.¹⁶⁷

Also, as has been seen, transparency is just one prong of a multi-faceted approach to the regulation of chemical substances. Restrictions may be imposed on the manufacture, placing on the market, or use of substances, including in articles. Already Annex XVIII includes restrictions designed to protect the general public, including through a range of prohibitions on sale to it. Requirements to provide information, including to consumers, do not appear to be a regular feature of the current restrictions laid down.

Moreover, substances of very high concern are subject to authorization, including for use in articles. Nonetheless, as outlined above, even for the most dangerous substances authorization may be granted where socio-economic benefit is deemed to outweigh risk, and where there is no suitable, technically and economically feasible, alternative. Thus even substances acknowledged to be of very high concern will remain on the marketplace, including for inclusion in consumer products. It is true that authorization may be made subject to conditions, and that there is no reason why the provision of consumer information could not form the basis of any such conditions. However, this remains an ad hoc, uncertain, case-by-case solution, to what seems to be a systemic weakness in the Regulation.

There are a number of reasons to think that REACH would be improved by incorporating more extensive requirements for the provision of information to consumers about substances in articles, especially substances of very high concern. This would enable consumers to make independent, informed, judgements about risk, in respect of substances over which there is uncertainty and disagreement among experts. It would also provide a push toward the substitution of dangerous chemicals by less risky alternatives, harnessing the power of the market to generate incentives for research and development, and for recourse to safer alternatives.¹⁶⁸

---

¹⁶⁶ Recall though that when authorizing a substance, including for use in an article, conditions can be imposed and these could also take the form of consumer information requirements. Likewise, restrictions may limit the sale of articles containing hazardous substances to the general public. There are many examples of this in the existing Annex XVII.

¹⁶⁷ REACH Implementation Project (RIP) Working Group 3.8. For general information on RIP, and for the Guidance on the Requirements for Substances in Articles, see: <http://ecb.jrc.it/reach/rip/>. For the kind of debate this has stimulated see, for example, the submission by Edana (the International Association Serving the Nonwovens and Related Industries) concerning the status of impregnated tissue paper as an article rather than as a container with preparations. See <http://www.europeantissue.com/Files/070307-EDANA+%20ETS%20Paper%20on%20Articles%20under%20the%20REACH%20Regulation.pdf>.

¹⁶⁸ This argument gains strength when we look to experience in California in relation to 'Proposition 65'. See Applegate, supra n. 4, for an overview and further references. He observes that '[a]s California has found with its Proposition 65, pointed public information is a strong incentive to use only the safest chemicals'.

C. Access to Information

The Chemicals Agency will be subject to the general EU law framework on access to information.[169] REACH, however, is proactive in indicating the accessibility of different categories of information. To this end, it establishes a traffic lights approach, laying down three lists which we call red, amber, and green.[170] Information included in the red list will not normally be accessible; it 'normally' being deemed to undermine the protection of commercial interests.[171] This includes details on the full composition of a preparation, the precise use of a substance or preparation, except in so far as this is required as part of the registration or authorization process, the precise tonnage of the substance or preparation, and information pertaining to links between a manufacturer or importer and his distributors or down-stream users. The Agency may, however, disclose this information where urgent action is essential to protect human health, safety, or the environment, including in emergency situations.[172] Information in the amber list will be available, except where a party submitting the information provides a justification at the time of registration as to why publication could be harmful to its or another party's commercial interests, and where this justification is accepted as valid by the Agency.[173] Information on the green list will be made publicly available.[174]

The categories of information listed on both the green and amber lists extend to information about substances on their own, as well as those included in preparations and in articles. On this basis, consumers will, invariably,[175] be able to obtain upon request to the Agency, information about the intrinsic properties of substances, including those in articles, details of human health and environmental thresholds (DNELs and PNECs), information about classification and labelling, and guidance on safe use. More often than not, and absent a justification to the contrary, they will also be able to receive summaries of tests and other activities undertaken to generate information for registration purposes, and information in the SDS.[176] All of this shall be made publicly available, free of charge, over the Internet.

Given the focus on substances in articles above, it is important to consider how much information the Agency will hold on these as a result of the other dimensions of REACH. First, it will hold information on substances in articles where these must be registered.[177] Second, the Agency will hold a more limited

---

[169] Article 118 explicitly provides that Regulation 1049/2001 applies to it.
[170] Articles 118–119. [171] Article 118(1).
[172] Article 118(2). [173] Article 119(2).
[174] Article 119. [175] Article 119(1).
[176] Article 119(2).
[177] Recall Article 7(1) and (5), namely when the one tonne threshold is met and the substance is intended to be released under normal or reasonably foreseeable conditions of use, or the Agency suspects that it will be, thus presenting a risk to human health or the environment.

amount of information on substances in articles which are notified to it; namely substances of very high concern, included on the candidate list for authorization, where the volume and concentration threshold is met.[178] Third, the Agency will hold information on substances in articles, where these substances appear on the final Annex XIV list specifying substances requiring authorization.[179] Given the information to be supplied to the Agency under these various headings, it is apparent that the vigorous consumer would be able to ascertain which substances are used in articles, at least where the release condition is met or where the article appears on either the candidate or final list for authorization. For substances in articles requiring registration or authorization, copious other information on that substance will also be available. This is especially true of substances requiring authorization, which will entail the supply of a CSR and an analysis of alternatives. The availability of information on alternatives is of the utmost importance, as it will allow consumers and non-governmental actors to target their campaigns to situations in which substitution is feasible but economically costly.

Bringing these three sections on access to information together, especially in relation to information on substances in articles, there would seem to be a missing link. While the Agency will hold considerable information, and in the main give public access to this, neither the supply-chain information requirements, nor the labelling rules, are such as to alert consumers to the presence of substances in articles, even where these substances appear on the candidate or final list for authorization, and are thus acknowledged to be of very high concern. It would be an important step in the right direction, were the labelling rules to require the presence of such substances in articles to be highlighted on a label. Even if it were not required that the substances present be named individually on the label, it would be an important step to acknowledge the presence of such substances as a category and to provide details of where consumers might turn to receive further information.

7. Conclusion

The REACH package is complex and rich in its governance dimensions. Even in relation to these dimensions, this essay is far from being comprehensive. Other arrangements for collaboration between Member States in the implementation of REACH, including arrangements for networked norm elaboration through soft law, have not been examined here.[180] These arrangements, as well as the work of

---

[178] Article 7(2).
[179] Recall Article 62 identifying the categories of information to be submitted as part of the applications process.
[180] See supra n. 167. See Lee, M. and von Homeyer, I. in current volume for a discussion of the comparable process under the Water Framework Directive. Thus, for all the differences between REACH, with its emphasis on harmonization, and the Water Framework Directive, with its

the Agency Forum briefly touched upon above, require further investigation, to consider their contribution to effective and/or legitimate transnational governance in this area.

Even without these aspects though, the REACH package is interesting. It is about product market integration *and* environmental/health protection. While the former militates in the direction of harmonization, the latter militates against a complacent centre, convinced that it enjoys a monopoly on regulatory wisdom. In keeping with this, REACH is emphatic but tentative in the harmonization which it achieves. It is emphatic in that it leaves little room for unilateral Member State departure from it, as is exemplified by the free movement clause. At the same time, and in a manner which might seem contradictory but is not, REACH is tentative in the harmonization which it achieves. It constructs multiple channels which allow for the contestation of the regulatory bargain which it embodies. Far from being complacent, the EU centre actively institutes a multi-level and multi-actor dialogue about risk and about how best to respond to it.

emphasis upon appropriate flexibility, each of these has spawned a broadly similar process for norm elaboration. In the Water Framework Directive this is known as the Common Implementation Strategy, in relation to REACH (humorously or not) it goes by the acronym 'RIP', meaning the REACH Implementation Projects. See: <http://reach.jrc.it/> for an insight into the many activities being conducted under this rubric.

4

Building Spatial Europe: An Environmental Justice Perspective[1]

Dr Jane Holder

1. Introduction

The European Spatial Development Perspective (ESDP)[2] sets out a framework for decision-making about land use and development in the European Union. Its main aim is to give planners and policy makers in the Member States a European reference point when they are drawing up development plans, applying for assistance from the EU, or giving consent for individual housing, transport, commercial, or energy projects.[3] The idea is that decisions about development should be made with respect to their impact on the 'Territory of the EU', and in turn the ways in which the global economic competitiveness of this 'Territory' can be enhanced. This requires the 'capacity to conceptualise or think about one's location or situation within the spatial structure of Europe as a whole',[4] with the ESDP providing a 'mental map' to help with this 'spatial positioning'.[5] Adopted a decade ago, the ESDP represented the EU's first attempt to coordinate and integrate social, economic, and environmental policies from a spatial perspective and under a sustainable development 'banner'. It was initially presented as bringing about an important next stage in (deeper) European integration, but has

---

[1] Many thanks to Katherine Lake and Denise Daly for research assistance on spatial strategy, and to Donald McGillivray, Joanne Scott and Ian Bache for their helpful comments on the paper. Thanks also to the participants of the *Modern Law Review* Seminar on 'Seeking Solidarity in the European Union—Towards Social Citizenship and a European Welfare State?', University of Sussex, May 2008, for their invaluable comments on the territorial solidarity aspects of the paper.
[2] Committee on Spatial Development, *European Spatial Development Perspective—Towards Balanced and Sustainable Development of the Territory of the European Union (ESDP)*, Presented at the Informal Meeting of Ministers Responsible for Spatial Planning of the Member States of the EU, Potsdam 10/11 May 1999 (CSD, 1999).
[3] ESDP, p. 11.
[4] Williams, R. H., *European Union Spatial Policy and Planning* (Paul Chapman, 1996), p. 97.
[5] Healey, 'The Place of "Europe" in Contemporary Spatial Strategy Making', *European Urban and Regional Studies* 5 (1998) 139.

equally been attributed with building European unity and identity by expressing certain common cultural values and a shared approach to spatial development.

In this chapter I give an account of the origins of the ESDP, its main objectives and some of its key elements—sustainable development, territory, mobility, environment, and cohesion. I highlight the function that the ESDP performs in reinforcing ideas of cohesiveness and solidarity in the EU, in particular by the use of the idea of a single 'Territory of the EU'. My main critique of the ESDP, and subsequent initiatives is the absence of any concern with environmental justice, or equal access to a clean environment and equal protection from possible harm irrespective of race, income, class, or any other differentiating feature of socio-economic status.[6] By comparing the EU's spatial strategy with the treatment of spatial issues in the United States, I suggest that the potentially 'hard edge' of such issues, such as the recognition of the unequal distribution of (to use the terminology of scholars in the United States) 'locally undesirable land uses' (or LULUs), has in the ESDP been overridden by concerns about the acceleration of economic growth in the regions and in rural areas, firmly in keeping with the original economic aims and objectives of the Union, but distanced from a more progressive environmental and social justice agenda.

This has very real consequences. Writing for the NGO, Capacity Global, Schwarte and Adebowale report that in many areas of Europe, but especially in central and eastern Europe, there are widespread environmental inequalities, creating 'potentially excellent model areas for environmental justice studies'.[7] The Coalition for Environmental Justice, a network of NGOs in central and eastern Europe, has compiled such a case study on settlements of Romani people, finding that many of these are sited near hazardous waste sites and former mines.[8] More generally the European Environment Agency's report on *Environment and Health* (2005) suggests that the environment-related share of the burden of disease is higher in lower-income countries and that climate change particularly affects vulnerable groups.[9] Still, Schwarte and Adebowale note, 'there is no systematic attempt within the EU to link social status, race or ethnicity to environmental risk exposure'.[10]

I suggest that the absence of any environmental justice content in the ESDP and subsequent spatial planning initiatives is because the conceptual framework

---

[6] Cutter, 'Race, Class and Environmental Justice', *Progress in Human Geography* 19/1 (1995) 1, cited by C. Schwarte and M. Adebowale, *Environmental Justice and Race Equality in the European Union* (Capacity Global, 2007). In this chapter, I am primarily concerned with spatial injustices, whilst recognizing the existence of other forms of environmental injustice e.g. the disproportionate use of water meters for poorer citizens.

[7] Schwarte and Adebowale, p. 16.

[8] See <http://www.cepl.ceu.hu>.

[9] European Environment Agency, Environment and Health, available at <http://reports.eea.europa.eu/eea_report_2005_10/en/EEA_report_10_2005.pdf>. See also Varga, Kiss, and Ember, 'The Lack of Environmental Justice in Central and Eastern Europe', *Environmental Health Perspectives*, 11 (2002) 110.

[10] Schwarte and Adebowale, supra, p. 18.

of the EU's spatial strategy is provided by a version of sustainable development which, whilst downplaying any welfarist or distributive justice content, supports an aggressive programme of advancing the place of the EU in the global economy. It is particularly important in this context that consultation processes which preceded the drawing up of the ESDP prioritized the views of professional, political, and corporate elites within Member States, and gave less emphasis to participation and deliberation at a local level which may have provided alternative or counter rationalities and discourses of development. This contrasts with the implementation phases of the ESDP, in which participation was enhanced because the shaping of decision-making 'on the ground' was considered vital for its effectiveness.

The ESDP's tenth anniversary seems a fitting point to assess its continuing legacy, particularly its ability to illustrate (perhaps exaggerate) in a spatial form many of the key dilemmas facing modern governance in the EU—the tension between centralization and decentralization, how to widen and deepen the political franchise in an expanded Union by securing active public participation and debate, and how to deal, simultaneously, with the regulatory demands associated with globalization and localization. My main concern, however, is with the tension between the pursuit of market-led solutions and active intervention in favour of environmental and social justice, a focus which leads me to the origins of spatial planning in Europe.

2. Bringing in Space in Europe[11]

The Council of Europe began work on a common spatial approach for the entire continent of Europe in the 1960s, and has since produced several documents along these lines.[12] Specific spatial policies began to be advanced by the EU in the late 1980s, albeit in an ad hoc manner, according to individual policy areas, most notably regional policy and transport policy, as key aspects of the achievement of European economic integration.

In terms of regional policy, securing evenly distributed economic growth has been a constant problem for the EU. There has long been a concern that 'regional inequality remains entrenched in Europe and seems to be growing rather than diminishing'.[13] Such regional disparity is matched by *intra*-regional differences, creating pockets of concentrated disadvantage, even in apparently prosperous

---

[11] A phrase taken from Healey, P., *Collaborative Planning: Shaping Places in Fragmented Societies* (Macmillan, 1997), p. 34.

[12] Council of Europe, *European Regional/Spatial Planning Charter* (the Torremolinos Charter) (CEMAT, 1983) and Council of Europe, *Guiding Principles for Sustainable Spatial Development of the European Continent* (CEMAT, 2000).

[13] Amin, A. and Tomaney, J., 'The Challenge of Cohesion' in Amin, A. and Tomaney, J. (eds.), *Behind the Myth of European Union* (Routledge, 1995).

cities. Such disparities were considered a major source of dissatisfaction in the period running up to the completion of the internal market, and thus capable of damaging the integration process by triggering a return to national protectionism.[14] In addition, criticism of the EU's approach to securing economic growth and greater cohesion through the market-led approaches which made up the project to complete the internal market led to a search for alternative strategies. One such strategy was the large-scale reform of the Structural Funds in 1988 which obliged regions to undertake a comprehensive system of regional planning in order to obtain financial support, and highlighted the necessity of the EU adopting a spatial perspective on cohesion.[15] The process of economic 'rationalization', including mergers, takeovers, and other activities designed to improve economic efficiency (supported in the main by 'non-spatial' EU policies such as competition policy), continued to form the basis of the drive to complete the internal market, but these bore most heavily on the regions, as economic activity became clustered around existing 'hot-spots'.

As the ESDP was being negotiated, a spatial perspective was in any event being built in to policy-making and decision-making in the EU via funding programmes. For example INTERREG, a (then) Community programme made under the European Regional Development Fund for economic development in less developed border regions, was introduced in 1996, to provide financial support for transnational projects such as flood and drought prevention. The framework for such assistance was later also seen as a means of implementing aspects of the ESDP, such as encouraging the cooperation of regional and local authorities in decision-making. This is now more strongly the case, with current INTERREG programmes (IIIB and IVB) requiring that recommendations made in the ESDP must be taken into account when decisions are made to fund certain transnational projects.[16] The close relationship between EU funding programmes and spatial considerations supports the suggestion that European spatial policy discourse initially relied heavily on EU regional policy, first for its 'institutional and economic leverage',[17] and then as a 'common carrier' of the ESDP's objectives into the Member States' planning and other policy systems.[18] The

---

[14] Giannakourou, G., 'Towards a European Spatial Planning Policy: Theoretical Dilemmas and Institutional Implications', *European Planning Studies*, 4/5 (1996) 595.

[15] See, e.g. Article 10 of the ERDF Regulation (Regulation 4253/88, OJ 1988 L 374/1) which encouraged the Commission to finance infrastructure projects and other projects having a marked Community interest. See further, Amin and Tomaney, supra, p. 18.

[16] CEC, *The EU Compendium of Spatial Planning Systems and Policies* (CEC, 2000) lays down guidelines for INTERREG III: trans-national cooperation (Strand B) proposals should 'build on the experience of INTERREG IIC and take account of Community policy priorities for TENS and of the recommendations for territorial development of the European Spatial Development Perspective (ESDP)'.

[17] Jensen, O., and Richardson, T., *Making European Space: Mobility, Power and Territorial Identity* (Routledge, 2004), p. 39.

[18] Roberts and Beresford, 'European Union Spatial Planning and Development Policy Implications for Strategic Planning in the UK' (2003) *Journal of Planning and Environmental Law* Supp (Occasional PA) 15, 18.

relationship of influence between the ESDP and regional policy is now more complex, with elements of the ESDP being absorbed within a broader, more integrated, concept of 'territorial cohesion'.[19]

In terms of disparities between Member States, the Cohesion Fund, established by the Maastricht Treaty, sought to mitigate the worst effects of the economic pressures which accompanied the move towards economic monetary union (EMU) by supporting environmental improvements and helping to develop trans-European networks in the poorest Member States. In 1996 policy guidelines set out 'a truly global vision'[20] of a network of road, rail, inland waterways, airports, and combined transport infrastructure which would reach into every corner of Europe, filling in the 'missing links' between national transport systems. The trans-European transport network (TEN-T) provides a good example of an early form of spatial planning, albeit one with a strong and almost singularly economic orientation, a point I return to below.[21]

As well as being seen by the Commission as 'a prerequisite for the acceleration and deepening of the process of economic integration',[22] the early evolution of a discrete spatial strategy at the EU level perhaps also inevitably arose from the 'hollowing out' of the nation state at this time,[23] triggered by a combination of globalization and regionalism, 'forcing the nation state simultaneously to concede power upwards to supra-national institutions... and downwards to local institutions with detailed knowledge of individual regions',[24] and creating pressure for transnational issues—including spatial issues such as transport and environmental protection—to be dealt with by international organizations, or on an intergovernmental basis, and under the banner of 'a Europe of regions'. The development of spatial planning then became part of a wider integration agenda in the EU, since regional, national, or Community projects in one country could have potentially negative impacts on the spatial structure (the land uses and landscapes) of other Member States, a recognition which led to relationships and partnerships being forged between regional governments and agencies and the Commission. This situation recalls the institutionalization of direct links between the Commission and non-central government actors such as regional and local authorities, local action groups, and businesses in the development of cohesion policy, so that the nation state was effectively bypassed in certain circumstances.[25] As Marks, Hooghe, and Blank describe: 'such links break open the mould of the

---

[19] See further, pp. 116–18 infra.
[20] Jensen and Richardson, supra, p. 7.
[21] See pp. 111–12.
[22] Giannakourou, supra, 602.
[23] See, on 'hollowing out' thesis generally, Rhodes, 'The Hollowing Out of the State: the Changing Nature of Public Service in Britain', *Politics Quarterly* 65/2 (1994) 138.
[24] Amin and Tomaney, supra, p. 37.
[25] Marks, Hooghe, and Blank, 'European Integration from the 1980s: State-Centric v. Multi-Level Governance' (1996) 34(3) *JCMS* 341.

state so that multi-level governance encompasses actors within as well as beyond existing states'.[26]

The main idea driving forward the ESDP project was that the integration of spatial factors into the implementation of Community policy at an early stage would help avoid regional disparities and additionally realign economic growth away from the core of commercial activity made up of London, Paris, Milan, Munich, and Hamburg.[27] The ESDP sought 'a spatial balance, designed to provide a more even geographical distribution of growth'[28] and, thereby, 'increased cohesion in the European territory'.[29] The following extract from the ESDP links the process of European integration with the need for integrated—or cross-sectoral—policy making:

The ESDP provides the possibility of widening the horizon beyond purely sectoral policy measures, to focus on the overall situation of the European territory and also take into account the development opportunities which arise for individual regions. New forms of co-operation proposed in the ESDP should, in future, contribute towards a co-operative setting up of sectoral policies—which up to now have been implemented independently—when they affect the same territory.[30]

Such statements provide a positive gloss on the process, but some have critically considered the influence of the economic and institutional properties of the European integration agenda on the development of a European level of planning policy. For example, Giannakourou has questioned whether the conceptual and ideological identity of such a policy can ever represent much other than a market-oriented integration system.[31] I further address this key question when considering elements of the ESDP, below.

Healey has usefully placed European spatial planning by the EU and Member States in historical perspective, reminding us that strategic planning in urban areas was widespread in Europe in the 1960s, a state-led movement she attributes to the need to manage urban growth, and promote welfare objectives such as housing, and encourage business efficiency.[32] By the early 1980s planned regulatory approaches to development had given way to 'project-led' planning, letting loose entrepreneurial and market forces in urban areas and creating ad hoc planning initiatives such as, in the United Kingdom, development corporations and planning 'holidays'. Healey examines how this shift advanced the acceptability of business concepts in planning, replacing the welfarist approach of planned space in the 1960s with a '... strategic promotion of place as a product in an external marketplace of investors, business relocators and tourists'.[33]

---

[26] Marks, Hooghe and Blank, supra, 369.
[27] This area in the centre of the EU including these metropolises has 40% of the EU's population, accounts for 50% of the EU's GDP and covers 20% of the EU territory (ESDP, p. 8).
[28] ESDP, p. 7.   [29] Ibid., p. 7.   [30] Ibid., p. 7.
[31] Giannakourou, supra, 602.
[32] Healey, supra, 139.
[33] Ibid., 140.

Giannakourou similarly identifies the fundamental concepts of spatial planning as originating in a discourse about welfare and redistributive justice.[34] She describes three theoretical pillars of justice at play in European policy: first, the concept of 'competitive spatial justice', promising the levelling of spatial imbalances through the redistribution of competitiveness among European areas; second, a concept of 'diversified spatial justice' tolerating discrimination of goals, instruments, and actors for the handling of divergent problems; and finally, a concept of 'pluralist spatial justice', appealing to both public and private stakeholders to contribute to the redistribution of spatial prosperity'.[35] For Giannakourou, these central ideas are a reorientation of a traditional concept of spatial justice, but in a new context of the competition principles of social and economic cohesion in a market-oriented spatial integration process; a reorientation which she considers profoundly paradoxical.[36] She further explains that although the first steps towards a European spatial policy might be mistaken as a step away from a liberal market territorial integration paradigm towards a market correcting planning approach, the formative shape of European level spatial planning policy far from replicates the standards of national welfare spatial policies—'[O]n the contrary, they are constrained by the dilemmas and ambiguities of the global European construction process.'[37]

In summary, spatial planning in Europe has its roots in state-led movements for welfare and justice. In the EU, the ESDP grew from major, but individual, practical initiatives in regional development, transport, and the completion of the internal market, all of which were directed towards firmly locating Europe at the heart of the global economy. Latterly, an overarching conceptual apparatus has been provided by sustainable development, creating a more coherent, but also more controversial, framework for spatial development in the shape of the ESDP.

A. The European Spatial Development Perspective

1. Impetus, origins, and legal base

Until recently the European Union had no dedicated spatial policy and in many cases the objectives of its policies—at least as defined in the Treaties—had no explicitly spatial character. And yet, it has long been clear that many of the EU's key policies have profoundly (and often adversely) altered land use patterns and landscapes in the Member States—agricultural policy, environmental policy, transport policy, regional policy, energy policy, and cohesion policy being the main examples. Recognition of this contradiction began a decade long process of developing an EU spatial strategy culminating in the ESDP. Rivolin describes how, lacking a legal base in the Treaties, the Commission simply played a proactive role in initiating the ESDP process, but that it ended its support as soon as

---

[34] Giannakourou, supra, 601.   [35] Ibid., 603.   [36] Ibid., 602.
[37] Ibid., 602.

the Member States became engaged in the process.[38] When that happened, Faludi explains, planners from one Member State holding the EU presidency after another took the lead in the ESDP process, 'putting their stamp on the proceedings and handing the draft over to the next one',[39] until under the German presidency of 1999 everyone agreed on the final version.

During these negotiations, the EU had no formal competence in this area, despite an early reference to spatial planning in the EC Treaty[40] and the argument that the EU had some residual competence in this area because it concerned the coordination of existing Community policies which affect the use, organization, and structure of the EU territory.[41] Whilst the lack of a specific legal base did not unduly restrict the development of regional or environmental policy,[42] spatial planning proved more controversial which accounts for the informal nature of the negotiations leading up to the ESDP and the unclear demarcation of responsibility between Member States and representatives of the Commission: the ESDP was prepared by an ad hoc committee (the Committee on Spatial Development), operating under the authority of ministers responsible for spatial planning in the Member States and styled as an 'informal council', meaning that it was unable to take any formal decisions, least of all exercise any delegated powers. Faludi maintains that the administrative machinery was kept deliberately informal in this way 'to keep it out of the clutches of the Commission';[43] 'the Commission' here referring to a handful of officials at the Directorate General for Regional Development (DG Regio) involved in the process for reasons of 'bureaupolitical *raison d'etre*', or 'the assumption that in the fullness of time responsibility for spatial planning would come their way'.[44] When the ESDP was finally agreed, the Commission dismantled the CSD, moving the agenda to a newly established

---

[38] Rivolin, 'The Future of the ESDP in the Framework of Territorial Cohesion' (2005) 2 disP 19.

[39] Faludi, A., 'The Application of the European Spatial Development Perspective', *Town Planning Review* 74/1 (2003) 1, 2.

[40] Article 130s EC mentioned spatial planning as an instrument for the accomplishment of the Community's environmental policy objectives (prudent and rational use of natural resources and preserving, protecting and improving the environment'. If ratified, the Treaty of Lisbon will amend the Treaty on European Union so that Article 3(3) will state that the objectives of the Union include promoting 'economic, social *and* territorial cohesion, and solidarity between Member States' (emphasis added). The Treaty of Lisbon also provides that territorial cohesion is a shared competence between the Union and Member States. See also possible amendments to Article 158 EC adding *territorial* cohesion, and providing a new legal base for this (new Article 174 Treaty Establishing the European Union) and revised Protocol on Economic, Social and Territorial Cohesion. 'Territorial cohesion' is the new term of art for spatial planning, discussed further pp. 116–18, infra). See further on competence, Faludi, 'The Learning Machine: European Integration in the Planning Mirror' (2007) *Environment and Planning* 2 (2007).' and on territorial cohesion, Rivolin, n. 38, supra.

[41] Bastrup-Birk and Doucet, 'European Spatial Planning from the Heart', *Built Environment* 23/4 (1997) 307, 312.

[42] Regional policy and environmental policy were developed long before being formalized by their inclusion in the EC Treaty by the Single European Act. See further Faludi, 'The ESDP: Shaping the Agenda', *European Journal of Spatial Development* 21 (2006) 1, 4.

[43] Faludi, 'The ESDP: Shaping the Agenda', 4.

[44] Ibid., 4.

subcommittee of the Committee on the Development and Conversion of Regions, known as the Committee on Spatial and Urban Development (SUD) with a Commission representative in the chair, and thus nearer to the policy home of 'territorial cohesion', in anticipation of a firm legal base for this being provided in the near future in the form of the Constitutional Treaty.[45]

The lengthy process of drawing up the ESDP provides a fine example of network governance at the EU level, in which a diverse group of national spatial planning policy-makers came to form an 'epistemic community' around the concepts and ideas of a European planning perspective.[46] Bohme describes this 'story of a group of national actors coming together again and again, and in the fullness of time forming what further down will be described as a policy network to establish a European planning discourse'[47] as a form of 'discursive European integration'.[48] From an environmental justice perspective, a vital issue is the lack of public participation in such a 'bottom-up' initiative,[49] deemed to be 'the result of a Europe-wide process of public debate'[50] (a claim I return to when discussing the absence of an environmental justice dimension to the ESDP and subsequent spatial planning initiatives[51]). The final document, produced by the Committee on Spatial Development and agreed (not, notably, adopted) by the Council of Ministers at Potsdam in 1999, grandly 'conveys a vision of the future territory of the EU'.[52] More prosaically, it provides 'a general source of reference for actions with a spatial impact, taken by public and private decision-makers'.[53]

With the lack of a legal base for spatial planning in the EU Treaties, the ESDP's legal effects remain questionable: as an inter-governmental, non-binding, instrument, without even the status of a policy, plan, or soft law, it serves 'only' as a framework for decision-making by the Member States, their regions and local authorities, and the European Commission, each acting 'in their own respective spheres of responsibility'.[54] However, as Faludi states, the provision of a 'frame of reference' can make a significant contribution, albeit one that is difficult to quantify in terms of 'hard results':

'Framing is what frameworks do—injecting ideas into the proceedings, ordering thoughts and thereby, albeit indirectly, giving direction to action. In so doing, the power of

---

[45] Ibid., 6.
[46] Faludi, Zonneweld, and Waterhout, 'The Committee on Spatial Development: Formulating a Spatial Perspective in an Institutional Vacuum', in Christiansen, T. and Kirchner, E. (eds.) *Committee Governance in the European Union* (Manchester University Press, 2000).
[47] Bohme, 'Discursive European Integration: the Case of Nordic Spatial Planning', *Town Planning Review* 74/1 (2003) 11.
[48] Ibid., 11.
[49] Faludi, Zonneweld and Waterhout, supra, n. 46.
[50] ESDP, p. 12.
[51] See infra, pp. 123–4.
[52] ESDP, p. 11.
[53] Ibid.
[54] ESDP, Preface.

frameworks must not be underestimated. However, frameworks do not impose themselves. Rather, they work on the minds of those who take its message into consideration. This, then is what 'application' [of the ESDP] means—ideas stimulating future action to take a particular course, but without pre-empting the decisions of those involved'.[55]

Despite uncertainties about its precise legal effect, the ESDP clearly anticipates in the longer term a broader EU level of planning activity[56] (this is almost certainly to be the case should the Lisbon Treaty be ratified (with this giving 'territorial cohesion' a firm legal base)).[57] And, in spite of its apparent lack of legal teeth, the ESDP is increasingly guiding European funding and influencing planning activity across Europe—as mentioned above, the INTERREG programme (which at first provided a framework for considering spatial issues in decisions about the distribution of funds) has since the 1990s proved to be a de facto mechanism for implementing elements of the ESDP.[58] Furthermore, the ESDP's guiding principles—sustainable development and policy integration in decision-making—now have firm legal underpinnings, having undergone a mainstreaming process in the EC Treaty.[59] So, although the ESDP itself has no enforceable legal effects, its aims, content, and method are shaped by law, and, as I discuss further below, it has proved capable of influencing EU and Member State governance of a wide range of issues. The ESDP has also triggered further inter-governmental initiatives designed to update its message, most notably the linked inter-governmental reports of 2007, *Territorial State and Perspectives of the European Union* (a comprehensive and lengthy ESDP-like document) and the truncated *Territorial Agenda of the European Union*[60] (based on the *Territorial State* report and amounting to an action plan which committed ministers to further developing territorial cohesion policy at the inter-governmental level prior to the signing of the Lisbon Treaty which added territorial cohesion to the existing legal base of 'social and economic cohesion').

---

[55] Faludi, 'The Application of the European Spatial Development Perspective', 2.
[56] Jensen and Richardson, supra, p. 21.
[57] See n. 40, supra.
[58] Jensen and Richardson, supra, p. 22.
[59] Articles 2 and 6 EC.
[60] German Presidency, *Territorial State and Perspectives of the European Union: Towards a Stronger Territorial Cohesion in the Light of the Lisbon and Gothenburg Ambitions* (2007) <http://www.bmvbs.de/Anlage/original_1005296/The-Territorial-State-and-Perspectives-of-the-European-Union.pdf>; German Presidency, *Territorial Agenda of the European Union: Towards a More Competitive and Sustainable Europe of Diverse Regions* (2007) <http://www.bmvbs.de/Anlage/original_1005295/Territorial-Agenda-of-the-European-Union-Agreed-on-25-May-2007-accessible.pdf>. These documents were preceded by Luxembourg Presidency, *Scoping Document and Summary of Political Messages for an Assessment of the Territorial State and Perspectives of the European Union: Towards a Stronger Territorial Cohesion in the Light of the Lisbon and Gothenburg Ambitions* (2005) (the first working draft of the *Territorial State and Perspectives* document) <http://www.eu2005.lu/en/actualites/documents_travail/2005/05/20regio/Min_DOC_1_fin.pdf>.

2. Key objectives

The orientation of the ESDP is strongly economic. Participation in and access to the global market by all areas in the European territory are its overriding justifications. The main aim is *polycentric* development, the spatially balanced distribution of economically active areas, capable of contributing to the global economy (polycentricity is presented in the ESDP by the metaphor of a 'bunch of grapes', replacing, variously, the 'blue banana', the 'Dorsale Europeenne'—the 'backbone of Europe'—and most recently the 'pentagon', as graphic representations of concentrated economic activity at the 'core' of the EU). Whereas previous plans for regional development had focused upon improving transport links between the periphery and the core, the ESDP seeks to achieve polycentricity by encouraging the creation of several 'dynamic zones of global economic integration', well distributed throughout the EU territory and comprising a network of internationally accessible metropolitan regions and their linked hinterland (towns, cities, and rural areas of varying sizes).[61]

> The economic potential of all regions of the EU can only be utilised through the further development of a more polycentric European settlement structure. The greater competitiveness of the EU on a global scale demands a stronger integration of the European regions in the global economy.... The creation and enlargement of several dynamic global economy integration zones provides an important instrument for the acceleration of economic growth and job creation in the EU, particularly also in the regions currently regarded as structurally weak.[62]

According to the ESDP, polycentricity is to be achieved by the development of 'functional regions' offering gateways to sea ports and airports, the construction of 'Euro corridors' to connect cities (or 'nodes' in the polycentric system), and cooperation between cities on matters of economic competition, culture, and education through cross-border and transnational networks.[63] The full participation of the 'lagging', or less economically favoured, regions in the global market is to be secured by improving their environment, service infrastructure, and access to knowledge and infrastructure, thereby increasing their attractiveness to investors.[64] This also relates to a key aim of reducing disparities between urban and rural areas (overcoming the 'outmoded dualism between town and country'),[65] by encouraging economic activity in the predominantly rural regions, for example fostering eco-tourism, and securing high speed transport links. The ESDP cites as 'a promising approach' to such decentralized development the Scottish Highlands, where small and medium-sized enterprises have obtained access to information and communication technologies with government support and can therefore 'tap into global markets'.[66] However, such a functionalist

---

[61] ESDP, p. 20. [62] Ibid., p. 20. [63] Ibid., p. 21.
[64] Ibid., p. 26. [65] Ibid., p. 19. [66] Ibid., p. 67.

emphasis on the role of the regions in the global economy has raised concerns about the treatment of rural areas under the ESDP.[67]

The ESDP is now ten years old. Although described by some as 'old hat',[68] it has not been surpassed and remains influential in law and policy (although this influence has been described as 'indirect and implicit rather than direct and explicit in nature')[69] with the current, post-ESDP process[70] being shaped by the original document and broadly in accordance with its aims and methods. For example, the inter-governmental reports of 2007, *Territorial State and Perspectives*, and *Territorial Agenda*, remain faithful to the ESDP's key objective of securing European competitiveness (in the light of the Lisbon and Gothenburg ambitions to become 'the most competitive and dynamic knowledge-based economy in the world, capable of sustainable economic growth with more and better jobs and greater social cohesion'). These reports borrow the ESDP's key concepts and 'vision' (a certain idée) of spatial planning in the territory of the EU, whilst also updating the challenges facing the Union and setting out the new concepts of 'territorial capital', 'territorial integration', and 'territorial solidarity', each of which are discussed further below. In summary, the ESDP set an important precedent for European spatial planning; this precedent has been further elaborated and distinguished according to new challenges and conflicts facing the Union, and in the light of further research, but with the ESDP's basic principles and message remaining intact.

In expressing these objectives, the ESDP has created a new policy language, supported and communicated by a new (and still shifting) vocabulary, an array of acronyms, symbols, metaphors, maps, texts,[71] and 'visions', and implemented by new 'knowledge building' institutional structures and the setting up of policy networks. A small but lively body of critical scholarship has also grown up around the ESDP.[72] Much of this work centres on the idea that the ESDP originated and

---

[67] On the general focus on competitiveness in the ESDP, see Richardson, 'Discourses of Rurality in EU Spatial Policy: The European Spatial Development Perspective', *Sociologia Ruralis* 40/1 (2000) 53.

[68] Faludi, 'The Learning Machine'.

[69] European Spatial Observatory Network (ESPON), *ESPON Project 2.3.1. Application and the Effects of the ESDP in the Member States: Final Report* (2007). <http://www.espon.eu/mmp/online/website/content/projects/243/366/index_EN.html>.

[70] On the post-ESDP process, Faludi, 'The King is Dead—Long Live the King! Why There Is No Renewed European Spatial Development Perspective and What Happens to the ESDP Agenda Anyhow?' Paper given at the Regional Studies Association International Conference, 'Shaping EU Regional Policy: Economic, Social and Political Pressures', 2006 <http://www.regional-studies-assoc.ac.uk/events/leuven06/Faludi.pdf>.

[71] The ESDP Action Programme, agreed by the Spatial Planning Ministers at Tampere, Oct. 1999, mandates the creation of a new textbook intended to form the basis of a common education about EU geography.

[72] See, for example, Faludi, A. and Waterhout, B., *The Making of the European Spatial Development Perspective: No Masterplan* (Routledge, 2002); Jensen and Richardson, supra, n.17; Pedrazzini, L. (ed), *The Process of Territorial Cohesion in Europe* (FrancoAngeli DIAP, 2006).

continues to operate as a mandate for economic growth and competitiveness, signalling the beginning of a critical geography of law, land, and power in the EU.

3. Governance

The ESDP envisages the use of a range of by-now familiar 'new' governance tools: the setting up of information-exchange structures, requiring the European Commission and Member States to agree upon spatial criteria and indicators, and the undertaking of long-term research, studies, and pilot projects by spatial research institutes in the Member States. Inclusive decision-making is considered particularly important, and, in the context of the structural funds, partnership arrangements are encouraged 'to mobilise...all relevant regional players in the decision-making process'.[73] Impact assessment features as a key part of the decision-making machinery, or at least a new form of this[74]—territorial impact assessment—described by the ESDP as a basic prerequisite for all large transport projects.[75] Elsewhere, the ESDP is considered to provide the initial basis for assessment (even though the statistical data on impacts and demographic changes is at a fairly basic level).[76] It is networks, however, which have emerged as the dominant method of spatial governance, an example of which is the setting up of the European Spatial Planning Observatory Network (ESPON) to work with the national institutes and the CSD (and its successor) with the aim of nourishing the discussion and policy development related to territorial development and cohesion by gathering information about territorial structures, trends, and policy impacts in the EU.[77] For instance, ESPON's 2007 project on the application and effects of the ESDP brought together various 'elite' groups: members of AESOP (Association of European Schools of Planning), planners and experts from accession countries, European Commission officials, and other actors in European planning on national and regional levels.[78] The most recent elaboration of the ESDP, the *Territorial Agenda*, refers to such examples of 'territorial governance' as producing 'an intensive and continuous dialogue between all stakeholders of territorial development'.[79] Generally though, the focus has been on the

---

[73] ESDP, p. 16.

[74] Originating in environmental policy, impact assessment is now a key feature of many different policy contexts, see further Scott and Holder, 'Law and New Environmental Governance in the European Union', in De Burca, G. and Scott, J. (eds.) *Law and New Governance in the EU and the US* (Hart, 2006).

[75] ESDP, p. 26. See further Williams, Connolly, and Healy, 'Territorial Impact Assessment: A Scoping Study', Final draft submission to the Committee on Spatial Development' (ESPRIN Study Team, 2000).

[76] ESDP, p. 14.

[77] Ibid., pp. 37–38. A further example of the network approach in this area is the TERRA programme, an 'experimental laboratory in spatial planning' (see CEC, *TERRA* (CEC, 2000) which falls under Article 10 of the ERDF Regulation).

[78] ESPON, *ESPON Project 2.3.1. Application and Effects of the ESDP in the Member States.*

[79] *Territorial Agenda*, p. 2. More generally on European network governance in spatial planning, see Bohme, 'Discursive European Integration', *Town Planning Review* 74/1 (2003) 11.

creation of high level forums for policy impact research, with less attention given to capacity building at local and regional levels during negotiation of the ESDP and afterwards.

The ESDP (and post-ESDP) approach to governance is multi-level and intergovernmental, but also allied to the working of Commission committees, in this case the CSD and its successor. The questions about EU competence over spatial planning issues appear in the case of the ESDP to have created the conditions for novel, informal, and cooperative methods of governance, including the development of new and complex policy webs, and 'knowledge-building' and 'knowledge-diffusing' institutions, all currently operating in 'a no-man's land outside the EU treaties'.[80] According to Faludi, this amounts to a 'system-like governance form in the making, albeit without firm and well-institutionalised principles of regulation in place'.[81] On the other hand, Giannakourou argues that the voluntary, cooperative, and horizontal nature of the spatial experiment of the ESDP does not reflect an undeveloped system-in-waiting, but instead a fully-fledged model of coordinated European or transnational spatial *options,* slowly replacing vertical and centralized forms of Member State development planning policy, and aiming at the dissemination of knowledge and successful practice on spatial issues, and a change in thinking brought about by the 'interpenetration of different institutional, administrative and cultural traditions'.[82] In line with this governance style and objectives, the ESDP provides general frames of reference and action, proposed standards of good practice and recommendations, all 'expressive of information',[83] rather than setting substantive rules and legally enforceable objectives and action.

B. Elements

As suggested by its description as a 'biblical text for European space',[84] the scope of the ESDP is considerable, covering transport, energy, natural and cultural heritage, information technology, structural funds, employment, water resources, cultural landscapes, environmental protection, nature conservation, agriculture and forestry, demographic changes, and enlargement of the EU. To give some meaningful sense of this, I address in greater detail several of its key elements—sustainable development, territory, mobility, environment, and cohesion—in particular by drawing upon the work and opinion of critical geographers and planners. I also indicate the points of influence and interrelationships between these elements, before offering a legal critique of the ESDP from an

---

[80] Faludi, 'The Learning Machine', 3.
[81] Jensen and Richardson, supra, p. 152.
[82] Giannakourou, supra, 607.
[83] Ibid., 617. See further, Streek, 'Neo-Voluntarism: A New European Social Policy Regime' (1995) 1 *European Law Journal*, 31, 40.
[84] Jensen and Richardson, supra, p. 8.

environmental justice perspective. Most importantly, as Dabinett and Richardson state, the emergence of the EU's spatial planning framework should *not* be 'understood as a purely comprehensive scientific rational process, or the benign convergence of national planning systems'.[85] Instead, analysis of these elements supports their argument that the ESDP bends to currently hegemonic ideologies of the single market and political integration, although admittedly also reflecting other debates about cohesion and the environment.

1. Sustainable development

The ESDP, avowedly modelled on the Brundtland Report's form of sustainable development, covers 'not only environmentally sound economic development which preserves present resources for use by future generations but also [includes] a balanced spatial development'[86] of the territory of the European Union'.[87] Three ambitious goals are to be achieved *equally* in all the regions of the EU: economic and social cohesion; conservation and management of natural resources and the cultural heritage; and more balanced competitiveness of the European territory. Sustainable development therefore threads together previously distinct policy areas into a conceptually coherent whole which seeks to balance the economic aims of previously separate but spatially important policy areas with the conservation and management of natural resources and the cultural heritage.

The ESDP provides what looks at first like an expanded form of sustainable development, so that the by now familiar inter-generational, or temporal, aspects of the principle are reinforced by the acknowledgement of existing spatial disparities and the wish to alleviate these. However, the ESDP recognizes these disparities almost entirely according to economic goals (rather than, for example, according to environmental justice criteria), in particular enhancing the competitiveness of Europe's 'peripheral', 'ultra-peripheral',[88] 'less favoured', or sparsely populated regions in the global economy. This, then, is a *particular* form of sustainable spatial development, one centred on a polycentric urban system, linked by transnational infrastructure networks, with a focus on the development of economic growth zones.[89] The 2007 Report, *Territorial State and Perspectives*, representing more recent inter-governmental thinking, similarly highlights the economic dimension of sustainable development: 'The key political challenge for the Union at this moment is to become economically more competitive and dynamic. Urgent action is needed if Europe wants to keep up its model for

---

[85] Dabinett and Richardson, 'The European Spatial Approach: the Role of Power and Knowledge in Strategic Planning and Policy Evaluation', *Evaluation* 5/2 (1999) 220, 228.l.
[86] ESDP, p. 9.
[87] ESDP, Preface.
[88] French Overseas Departments, Azores, Madeira, and the Canary Islands are all closely linked to other continents and are thus considered capable of giving the EU a head start in cooperation with their neighbouring countries (ESDP, p. 55).
[89] Jensen and Richardson, supra, p. 21.

sustainable development. This requires a stronger focus on growth and employment whilst also taking proper account of social and environmental issues.'[90] As mentioned, the 2007 *Territorial Agenda* uses as shorthand for such intentions the phrase 'sustainable growth'.[91] This is a subtle but significant change in emphasis: 'sustainable growth' had been dropped from the EC Treaty by the EU in favour of the less contentious (in environmental policy circles) 'sustainable development', with its allusion to forms of development other than economic.[92]

Whilst there is clearly scope for conflicts between the 'sustainability' goals set out in the ESDP and the post-ESDP documents, the very ideology of sustainable development is that such conflicts can be overcome. Working within a discourse of ecological modernization, the ESDP and subsequent documents appear to uphold that growth can be accommodated within existing social structures and environmental conditions and that there is no need to impose any limits on growth. The EU's form of sustainable development in its spatial planning strategy is therefore not only permissive of economic development, but is also narrowly focused, with little or no reference to the impacts of the EU's development patterns on the rest of the world. I return to this striking absence of any element of intragenerational justice when considering the relationship between environmental justice and spatiality in the EU.[93]

2. Territory

The Europeanization of state, regional, and urban planning is the stated objective of the ESDP,[94] presumably meaning in this context the penetration of a European spatial development perspective in the Member States. The use throughout the ESDP of the phrase 'The Territory of the EU'[95] supports this. However the very idea of a demarcated territorial identity for the EU is problematic because territory is so closely identified with the sovereignty of the nation state. The 'Territory of the EU' is presented by the ESDP as a given, a physical reality, whereas it is in fact a changing and uncertain thing. There is, for example, some recognition that the spatial development policy of the EU must extend beyond the territory of the Member States, considering the interests of neighbouring countries and cooperating with them.[96] 'The Territory of the EU' was also a partial affair at the time of the ESDP's agreement since countries which were negotiating their accession to the Union at this time were formally excluded from its scope. But there is some ambivalence towards their position,[97] with the ESDP proposing that Member States consider 'a new reference territory for the ESDP'[98]

---

[90] *Territorial State and Perspectives*, p. 7.
[91] Ibid., p. 3ff.   [92] Article 2 EC.   [93] See pp. 118–24.
[94] ESDP, p. 45.   [95] ESDP, p. 1ff.   [96] Ibid., p. 51.
[97] Negotiations had begun with Estonia, Poland, Slovenia, Czech Republic, Hungary, and Cyprus.
[98] ESDP, p. 46ff.

following enlargement and grant them pre-accession financial assistance (a policy direction which appears to have had some success as a means of applying the ESDP in the accession countries). These examples apart, the interests of Central and Eastern European countries appear to have been more fully expressed by the process of spatial planning undertaken by the Council of Europe.[99]

Healey has examined the use of ideas of territory in spatial planning exercises in Europe, including a tendency which involves 'a strategic rethinking of an urban region's economic, political and cultural position in relation to a range of potential external audiences'[100] and which, importantly, calls into play the metaphor of 'Europe', a sign of a 'marketing style' in spatial planning, or a form of entrepreneurial governance which positions urban region development planning in a 'European' territory:

> In this context, the metaphor of 'Europe' plays a key role in defining the wider space within which an urban region is positioned. But this positioning work is also accompanied by, and is a key component of, political mobilization efforts to build internal cohesion among urban region stakeholders. The dynamic behind these efforts arises partly from the need to respond to a new external economic and political environment. This environment is populated by diverse and often newly recognized stakeholders in urban region change, for whom a particular place has different meanings and assets, depending on their strategies and networks.[101]

Healey's analysis is based on examining spatial plans in 10 Member States, including, for example, the spatial plan designed to develop a conception of Lyons as a 'Eurocity' (or 'important economic and cultural node at the core of a supranational European space')[102] and hence capable of forging horizontal links with other cities in France and with similar cities elsewhere (thereby replacing the centrifugal relationship of power with the centre, Paris). In such examples, 'Europe' is clearly an important part of the spatial discourse, 'positioning' the urban area in a larger territory and connecting it with other cities and states. In cases such as this the EU as a political entity might be absent, but Healey's presentation of the uses of the 'Europe' metaphor in strategic plans at state and regional levels may be similarly applied to strategic planning initiatives at EU level. For example, returning to the ESDP, the need to create a single, economically cohesive, and productive bloc—the EU as a global player—is supported by the idea of the existence of the 'Territory of the EU', an 'imagined territory', reminiscent of Shore's 'imagined community' and representative of a common European culture and identity,[103] but which in this case gains additional potency from the more usual meaning of territory as a physical representation of

---

[99] Jensen and Richardson, supra, p. 23, referring to the Council of Europe's, 'Guiding Principles'.
[100] Healey, 'The Place of "Europe"', at 140.
[101] Ibid., 140.
[102] Ibid., 142.
[103] Shore, C., *Building Europe: The Cultural Politics of European Integration* (Routledge, 2000).

state, with its associated notions of unity and security. In this way, the ESDP can be seen as 'both articulating a functional network of regions and nation states in a competitive global region, as well as injecting a spatial dimension into the discourse of political integration in Europe, and thus potentially spatialising the less tangible notion of a European identity'.[104]

The idea of a single European territory in EU spatial planning (representing the EU as a unified political and, most importantly, economic, entity) further encourages a sense of solidarity and cohesion, as opposed to the vaguer appeal to 'Europe' in the plans studied by Healey. In practical terms, this means that decisions about development, at regional, Member State, and EU level should be made with reference to their impact on the 'Territory of the EU', and in turn the ways in which the global economic competitiveness of this 'Territory' can be enhanced. This process of 'spatial positioning' provides the main rationale for the ESDP, and is couched in terms of the 'Europeanization' of planning:

> It is proposed that Member States also take into consideration the European dimension of spatial development in adjusting national spatial development policies, plans, and reports. Here, the requirement for a 'Europeanization' of state, regional and urban planning' is increasingly evident. In their spatially relevant planning, local and regional government and administrative agencies should, therefore, overcome any insular way of looking at their territory and take into consideration European aspects and inter-dependencies right from the outset.[105]

Such clearly pronounced 'top-down' requirements are also combined with more informal exchanging and spreading of ideas through transnational structures and funding programmes, often involving cooperation between local government in different Member States, bypassing central government entirely, and thereby changing the power balance between nation states, regions, and local government.[106]

ESPON's Report on the application and effects of the ESDP in Member States concludes that planners have taken heed of the ESDP (especially those in the new Member States), rather less so those working in other policy sectors.[107] Within planning discourse, the effects relate to the ways in which the spatial representation of a country's place in a wider Europe—its 'spatial positioning'— should take place.[108] What ESPON reports as a strong degree of '*implicit* application' in the Member States, Faludi interprets as evidence that the ESDP is having 'learning effects', at least in respect of planners.[109] He finds, for example, that ideas and approaches percolate through and influence the way of thinking about planning in Europe. The result is:

> . . . a common approach for opening up and exploring new fields of European policy. The method is nowhere clearly postulated, but the prominence of stakeholder participation,

---

[104] Jensen and Richardson, supra, p. 46.
[105] ESDP, p. 45.
[106] Jensen and Richardson, supra, p. 181–2.
[107] ESPON, *ESPON 2.3.1*.
[108] Faludi, 'The Learning Machine', 12.
[109] Ibid., 10.

demonstration projects, fair distribution of projects and initiatives throughout the whole of Europe and the use of forums and so forth is obvious. So is the search for images and issues that transcend the borders of nation states... because they require a broader perspective... And, finally so is the predilection for concepts that are capable of uniting opposites, like polycentric development.[110]

To take the example of the United Kingdom, these 'learning effects' have been substantial, with planning legislation adopting the concept of integrated spatial planning, and local and regional levels of government enthusiastically taking on the idea of spatial positioning, even if only because of the likely positive consequences of aligning a project with current European thinking on territorial development for EU regional funding.[111]

The fiction of a single, cohesive, European Union territory therefore operates as the basic reference point for plan-making and decision-making. More elusively, it appears to have entered the 'spatial consciousness of urban region strategy makers'.[112] But the strategic emphasis upon a single European territory possibly leads to a certain downplaying of the intrinsic value of diverse and local land uses, particularly when this value is compared to the function of land uses from the perspective of European-wide spatial development. So, for example, the ESDP states that '[A] distinctive landscape can be used to promote the qualities of an area for attracting new industry, for tourism and for other types of economic investment.'[113] This sharply functional approach to nature, and natural resources, further contributes to the 'strategic promotion of place as a product',[114] as Healey describes it, capable of overriding the relationships between biological and social diversity and cultural identity.

An elaboration of this tendency can be seen in the 2007 *Territorial State and Perspectives* report which introduces into EU spatial planning strategy (renamed territorial cohesion policy in this report) a concept of 'territorial capital', developed originally by the OECD:

A region's territorial capital is 'distinct from other areas and is determined by many factors [which]... may include... geographical location, size, factor of production

---

[110] Faludi, 'The Application of the European Spatial Development Perspective', 8.

[111] Planning and Compulsory Purchase Act 2004, which introduced a form of integrated spatial planning in England and Wales. See also the thinking of the Royal Commission on Environmental Pollution, Report on *Environmental Planning*, Cm 5459 (HMSO, 2002) which was influenced by the ESDP and provided the impetus for the 2004 Act. Further discussed in Holder, J. and Lee, M., *Environmental Protection, Law and Policy*, (OUP, 2007) Ch. 13. On local and regional application of the ESDP, see discussion of case studies in ESPON, *Project 2.3.1. Application and Effects of the ESDP in the Member States*, pp. 156–79. See further, Tewdwr-Jones, 'Planning Modernised' (1998) *Journal of Planning and Environmental Law*, 519; and Shaw and Sykes, 'Investigating the Application of the European Spatial Development Perspective to Regional Planning in the United Kingdom', *Town Planning Review* 74/1 (2003) 31.

[112] Williams, R. H., *European Union Spatial Policy and Planning* (Paul Chapman, 2006), cited in Healey, 'The Place of "Europe" in Contemporary Spatial Strategy Making', *European Urban and Regional Studies* 5 (1998) 139, at 140.

[113] ESDP, p. 74.   [114] Healey, 'The Place of "Europe"', at 140.

endowment, climate, traditions, natural resources, quality of life or the agglomeration economies provided by its cities... Other factors may be 'untraded interdependencies' such as understandings, customs and informal rules that enable economic actors to work together under conditions of certainty, or the solidarity, mutual assistance and co-opting of ideas that often develop in small and medium sized enterprises working in the same sector (social capital). Lastly there is an intangible factor, 'something in the air', called 'the environment' and which is the outcome of a combination of institutions, rules, practices, producers, researchers and policy-makers, that make a certain creativity and innovation possible. This 'territorial capital' generates a higher return for certain kinds of investments than for others, since they are better suited to the area and use its assets and potential more effectively'.[115]

The EU's *Territorial State and Perspectives* report goes on to describe the exploitation of territorial capital 'as an important prerequisite for improving the global competitiveness of the whole EU territory'.[116] Spatial positioning is the key to achieving this: under the heading 'Strengthening Territorial Identity, Specialization and Positioning in Europe', regions are urged to 'identify within their territorial development policies their specific territorial advantages in an EU perspective'.[117] Cleverly, through the use of the territorial capital concept, the uniqueness and diverse character of areas are given an economic value, and also described in such terms. The development of this concept most likely represents an attempt to deal with the shortcomings of the ESDP's approach to decentralized economic growth (the 'bunch of grapes'), since polycentricity has been realized more by the expansion of the pentagon, than the development of alternative and networked 'dynamic zones of global economic integration'.[118]

'The Territory of the EU' is therefore a deeply contestable concept which performs an important function of advancing the *cohesive* effects of the 'Europeanization' of spatial planning. On the contrary, Jensen and Richardson consider that the device of a single European territory is also having a destabilizing effect. They explain that the emerging field of European spatial policy discourse is producing a new framework of spatialities—'of regions within Member States, transnational mega-regions, and the EU as a spatial entity',[119] which disrupts the traditional territorial order. Adopting Foucauldian analysis, they conclude: '[T]he new transnational orientation creates new territories of control, expressed through the new transnational spatial vision of polycentricity and mobility.'[120]

3. Mobility

The ESDP places a sustained emphasis upon the achievement of the trans-European transport network (TEN-T) as a means of delivering an efficient and

---

[115] OECD *Territorial Outlook* (OECD, 2001), p. 1.
[116] *Territorial State and Perspectives*, p. 3. [117] Ibid., p. 73.
[118] *Scenarios on the Territorial Future of the European Union*, p. 27.
[119] Jensen and Richardson, supra, p. 44.
[120] Ibid.

integrated transport system, capable of reaching into all parts of the European territory, and thereby 'connecting insular, landlocked and peripheral areas to the central areas',[121] and 'shrinking' Europe in the pursuit of the single internal market.[122] To this end, a focus of the ESDP is the development of high-speed railways in recognition that increases in traffic can no longer be managed by the expansion of road infrastructure alone,[123] but the 'great importance' of road traffic for both passengers and freight is still noted, especially for linking peripheral or sparsely populated regions'.[124]

The ESDP's disproportionate emphasis upon achieving the TEN-T supports an understanding of its rationale as a means of securing 'frictionless' free movement—of goods, peoples, services, and capital—throughout Europe.[125] To recap, the Treaty of Rome and its subsequent amendments put in place the machinery needed to dismantle legal barriers to trade, and in so doing provided the EU's core value of freedom of movement. The Single European Act seemingly swept away outstanding technical barriers. The ESDP turns its attention to the remaining physical barriers to trade which are capable of creating 'friction'—or delays. The solution is 'a transnational polycentric space connected by a single long-distance, seamless transport network',[126] first envisaged by the TEN-T, then absorbed and elaborated by the ESDP notably without having undergone any form of strategic environmental assessment (although this later became a requirement).[127] As Jensen and Richardson point out, the extensive infrastructure development required for the full implementation of TEN-T creates the single most difficult challenge to the achievement of the ESDP's vision of balanced, sustainable spatial development. However, because the concept of trans-European infrastructure networks became embedded in European policy and budgets long before the ESDP was drafted, spatial policy has been forced 'to swallow whole a programme which threatens to undermine its basic intentions'.[128]

The basic logic of the ESDP—that speeding up mobility and improving accessibility to the periphery and/or rural areas will lead to the creation of the single market *and* to balanced economic development of the Community—arguably rests upon a set of unproven assumptions about the effects of infrastructure development.[129] For example, new infrastructure development may

---

[121] ESDP, p. 14.
[122] Jensen and Richardson, supra, p. 53.
[123] ESDP, p. 14.
[124] Ibid., p. 28.
[125] Jensen and Richardson, supra, p. 20.
[126] Ibid.
[127] CEC, *TEN-T Policy Guidelines* (1996), Article 8: 'The Commission will develop methodology for Strategic Environmental Assessment of the entire network...', now overtaken by the requirements of Directive 2001/42/EC on the assessment of the effects of certain plans and programmes on the environment (the SEA Directive) OJ 2000 L 197, p. 30.
[128] Jensen and Richardson, supra, p. 21.
[129] Richardson, 'Trans-European Networks: Good News or Bad for Peripheral Regions', *Proceedings of Seminar A: Pan-European Transport Issues*, pp. 99–110 (23rd European Transport Forum, 1995).

bring remote regions 'closer' to the centre, in terms of travel time, but it may also create zones of relative peripherality, both in these regions and more generally. There is also the 'pump effect' problem of new high-speed infrastructure removing resources from structurally weaker and peripheral regions.[130] Whether producing the desired effects or not, the process of harmonization and promotion of mobility hints at a significant shift from a 'Europe of regions'—with an emphasis upon locality, 'situatedness', and 'place'—to a 'Europe of "flows"'. This is the term adopted by Hajer to describe the increasing reliance on low cost movement of goods by road, and a related demand for personal mobility,[131] resulting in sliced up urban and rural communities, ruined habitats and landscapes, and the loss of meaningful connections with 'places' and civil society. In sum, this is a picture of society and the environment being 'consumed' by infrastructure.[132] Richardson and Jensen further describe, albeit in exaggerated terms, the effect of this direction of European spatial planning as the creation of a European '*monotopia*'. This term captures the idea of a one-dimensional (mono) discourse of space and territory (topia/topos) which operates as an organizing set of ideas about the European Union territory according to a single overarching rationality of making a 'single European space' possible by the creation of networks enabling frictionless mobility—'an organised, ordered and totalised space of zero-friction and seamless logistic flows'.[133] They explain further:

...though the word 'monotopia' will not be found in any European plan, policy, document or political speech, this idea of monotopic Europe lies at the heart of the new ways of looking at European territory...a rationality of monotopia exists, and it is inextricably linked with a governmentality of Europe, expressed in a will to order space, to create a seamless and integrated space within the context of the European project, which is being pursued through the emerging field of European spatial policy.[134]

Such a 'will to order' European territory (with parallel processes taking place in relation to European skies and seas)[135] relies heavily upon ideas of cohesion and balanced, sustainable development, and the positive potential of a harmonized Europe, all of which tend to mask the reality that the planning of infrastructure programmes at the European level (as at national and regional levels) is strongly

---

[130] Ibid.
[131] Hajer, 'Transnational Networks as Transnational Policy Discourse: Some Observations on the Politics of Spatial Development in Europe', in Faludi, A. and Salet, W. (eds.) *The Revival of Strategic Planning* (Kluwer, 2000). The term was first coined by Castells, M., *The Information Age: Economy, Society and Culture, Vol 1: The Rise of the Network Society* (Blackwell, 1996).
[132] Jensen and Richardson, supra, p. 51.
[133] Ibid., p. 3.
[134] Ibid., p. 3.
[135] The 'Single European Sky' project seeks to create an integrated airspace over the territory (DG TREN, *Single European Sky: Report of the High Level Group* (Eur-OP, 2000); a process of marine spatial planning has also begun, e.g. see <http://ec.europa.eu/environment/water/marine/index_en.htm> on a marine strategy directive which aims to secure good ecological status for the marine environment by 2020, having regard in certain ways, to economic and social criteria.

influenced by powerful industrial lobbies, especially those representing road transport.[136]

4. Environment

The ESDP offers the potential to guide decision-making in an environmentally protective direction, recognizing, for example, the role spatial development policy can play in shifting road traffic to local public transport, and encouraging cycling and walking.[137] Renewable energies are described as 'particularly promising from a spatial perspective',[138] and common energy systems are considered as belonging to prudent environmental policy.[139] There is also a nod to urban ecology, with some reference to the conservation and development of small planted areas in urban green spaces,[140] and, in rural areas, recognition of the important role of European forests as 'green lungs'.[141] These are all well and good, except (as discussed above) the emphasis of the ESDP remains solidly on the future extension and improvement of existing infrastructure forming part of the TEN-T (albeit with some acknowledgment of the desirability of low impact transportation).[142] Such an emphasis rightly raises doubts about whether a form of urban-centred development which is so heavily reliant on increasing mobility can ever deliver environmental sustainability.[143] For example, the emission of carbon dioxide and other greenhouse gases as a result of European transportation patterns is barely addressed by the ESDP, beyond an introductory statement that the EU is 'a major contributor to world-wide CO_2 emissions together with the other large industrial countries and regions',[144] and a statement of faith in efficient transport systems and organization of settlement structures to reduce this contribution.[145] Notably, the ESDP offers very little sense of the EU's climate change impacts beyond the 'Territory of the EU', an approach replicated in the post-ESDP process.[146] That the impact of Europe on global climate change and other environmental dilemmas go unresolved in a European spatial policy centred upon road (and, increasingly, air) traffic has prompted the comment that 'the

---

[136] Jensen and Richardson, supra, p. 54. They give as an example the influential role of the European Round Table of Industrialists in setting the agenda for TENT-T policy by producing documents such as *Missing Links* (ERT, 1984). The ERT working group on infrastructure was chaired by Umberto Agnelli from Fiat, with Bosch, Daimler Benz, Petrofina, Pirelli, Total, and Volvo amongst the membership (Jensen and Richardson, supra, p. 73).
[137] ESDP, p. 14. See further Leibenath and Pallagst, 'Greening Europe? Environmental Issues in Spatial Planning Policies and Instruments', *Town Planning Review*, 17/1 (2003) 77.
[138] ESDP, p. 15.   [139] Ibid., p. 23.
[140] Ibid., p. 22.   [141] Ibid., p. 31.
[142] Ibid., p. 26.
[143] Jensen and Richardson, supra, pp. 91–2.
[144] ESDP, p. 10.
[145] Ibid.
[146] See, for example, ESPON, *Project 3.4.1. Europe in the World* (ESPON, 2006), although there is some reference to trade and inequality.

writers of EU spatial policy have been left with an unenviable task—of massaging a massive infrastructure programme and an unsustainable mobility scenario into a vision of sustainable spatial development'.[147]

In the ESDP's rearticulation of the territory of Europe, Jensen and Richardson identify a hierarchical ordering which places the environment as subsidiary to the logics of material growth and market expansion, so that: 'at stake in this globalised economic competition are biodiversity, environmental carrying capacity and landscapes of cultural heritage'.[148] By way of example, the ESDP takes a strongly economically oriented approach to the 'wise management' of the natural (and cultural) heritage. Whilst it is accepted that 'strict protection measures are sometimes justified' when the natural and cultural heritage of the EU is threatened, '... it is often more sensible to integrate protection and management of the endangered areas into spatial development strategies for larger areas',[149] in recognition that the natural heritage is an economic factor which plays an increasingly important role in the location decisions of new companies.[150] The ESDP also sees the need for less sensitive areas to be the subject of economic uses, presenting an unerringly functional approach to conservation:

Protection regulations and development restrictions should not be allowed to have a negative impact on the living condition of the population. Instead, ecological resources should be costed in economic terms—for instance through adapted fiscal solutions. Through earnings produced in this way, each region could open up appropriate new development opportunities, at the same time preserving the natural heritage.[151]

The main point here is that the value of the natural world is not merely subordinated to economic interests, but that this value has been subsumed by an economic rationality and methodology. In the ESDP, the economic 'pillar' of sustainable development prevails, leaving the fundamental tension between globalized economic competition and environmental protection resolved in favour of 'well distributed' opportunities for economic development throughout the European Territory, with the aim of contributing to social and economic cohesion.

5. *Cohesion*

As mentioned above, achieving cohesion provided an original rationale for the early development of a European perspective on spatial planning. However there was still some scepticism in DG Regio about the benefits of adopting a spatial approach to regional policy. Such concerns seemed to have been overcome by the time that the Second Report on Economic and Social Cohesion[152] was published.

---

[147] Jensen and Richardson, supra, p. 82.
[148] Ibid., p. 4.   [149] ESDP, p. 30.
[150] Ibid., p. 30.   [151] Ibid., p. 31.
[152] CEC, *Unity, Solidarity, Diversity for Europe, Its People and Its Territory: Second Report on Economic and Social Cohesion* (CEC, 2001).

This, as well as the Third Cohesion Report[153] and Community Strategic Guidelines for the Structural Funds,[154] were all heavily inspired by the ESDP, especially its core concept of polycentric development. In addition Member States' concerns about EU control of the planning and implementation of infrastructure projects making up the TEN-T (given the EU's lack of formal competence in spatial planning) led increasingly to the use of EU structural (and other) funds to secure the ESDP, and other spatial programmes, 'by the back door'.[155] Regional funding programmes in support of the aim of cohesion have since provided the main vehicle for the application of the ESDP in the Member States.[156] The linking of spatial and regional policy in this manner and also in institutional terms (the replacing of the CSD with a subcommittee falling within DG Regio's remit)[157] is indicated by the now widespread practice of renaming spatial planning as 'territorial cohesion', signalling a shift in the terms of the European spatial planning debate, in favour of a broader form of cohesion policy and (now) a firmer constitutional basis.[158] As the Commission explained in its Third Cohesion Report: '[T]he concept of territorial cohesion extends beyond the notion of economic and social cohesion by both adding to this and reinforcing it', in this case by reference to sectoral policies having spatial impacts (such as environmental policy) and other regional policies.[159] However, this process of extending the remit of cohesion policy, described as 'territorial integration',[160] is inevitably hampered by the current lack of indicators for the environmental and social dimensions of territorial cohesion, particularly when compared to the existence of established methodologies for measuring economic imbalances between regions in 'classical' (or the narrower form of) cohesion policy.[161]

In turn, the absorption of the ESDP's concepts and vision for the EU Territory in regional funding programmes has led to a greater focus on competitiveness in territorial cohesion policy. Whilst the Second Cohesion Report concentrated in the main on reducing disparities between regions ('Unity, Solidarity, Diversity for Europe'), by the Third Report cohesion is framed more in terms of development and competitiveness ('A New Partnership for Cohesion—Convergence,

---

[153] CEC, *A New Partnership for Cohesion—Convergence, Competitiveness, Cooperation: Third Report on Economic and Social Cohesion* (CEC, 2004).
[154] CEC, *Cohesion Policy in Support of Growth and Jobs: Community Strategic Guidelines 2007–2013*, OJ 2006 L 291/11.
[155] Jensen and Richardson, supra, p. 137.
[156] This is the key finding of ESPON, *Project 2.3.1.Application and Effects of the ESDP in the Member States*.
[157] See pp. 99–100, supra.
[158] Article 16 of the Draft Treaty Establishing a Constitution for Europe provided a legal base for territorial cohesion policy.
[159] CEC, *Third Cohesion Report*, pp. 57.
[160] German Presidency, *Territorial Agenda*, p. 1.
[161] ESPON, *ESPON Project 3.4.1. Europe in the World*, p. 113.

Competitiveness, Cooperation')[162]—an agenda which has been further pursued by the adoption of the Community Strategic Guidelines for the Structural Funds.

An important outcome of the post-ESDP process is the relabelling of the equity content of cohesion policy as *'territorial solidarity'*. This concept, as described in the *Territorial Agenda*, is concerned with securing 'better living conditions and quality of life with equal opportunities, oriented towards regional and local potentials, irrespective of where people live—whether in the European core or in the periphery'.[163] Although there is no clear definition of territorial solidarity, it seems primarily to describe opportunities to participate *equally* in economic life. The *Territorial Agenda* considers as an 'essential task and *act of solidarity*' to bring about 'equal opportunities for its citizens and development perspectives for entrepreneurship'.[164] The nature of these 'equal opportunities' is identified more clearly in Article 16 of the (now derailed) Draft Constitutional Treaty: '... citizens should have access to essential services, basic infrastructures and knowledge by highlighting the significance of services of general economic interest for promoting social and territorial cohesion'. The economically oriented nature of such equal opportunities suggests that territorial solidarity primarily carries more of a concern with the liberalization of services and freedom of movement than with social redistribution and the reconstitution of social rights at a European level, an emphasis sharply critiqued by Giannakourou (writing on early spatial planning policy):

Compared to national state welfarism, actual European-style spatial fairness follows a different conceptualization. In particular, while the former promises the redistribution of resources, services and incomes among the different areas of the national territory and, thus, the equalization of their developing conditions, the latter relies exclusively on the idea of equality of chances through the diffusion of information and knowledge on spatial questions to the national authorities and the market participants. Hence, what is attempted to be redistributed through the new European spatial strategy is not economic prosperity per se but just the chances to access economic prosperity and a better quality of life; in other words, the redistribution of competitiveness opportunities among European areas.[165]

Such analysis of the idea of 'equality of chances' supports a characterization of the EU's development of a spatial perspective as a market-accommodating policy, aimed at increasing mobility and competition among the European regions,[166] rather than as previously portrayed, a market-correcting policy concerned with the redistribution of public services and spatial justice in the European territory.[167]

---

[162] ESPON, *ESPON Project 2.3.1. Application and Effects of the ESDP in the Member States*, p. 113.
[163] *Territorial Agenda*, p. 1.
[164] German Presidency, *Territorial Agenda*, p. 3.
[165] Giannakourou, supra, 607.
[166] Ibid.
[167] Delors, 'The Social Policy of Europe', Communication to the European Conference of Brussels organized by the European Commission and the Department of European Law of the Catholic University of Louvain, May 1994, cited by Giannakourou, supra, 607.

In summary, there are now two separate processes of European territorial policy-making, both linked to the ESDP: inter-governmental cooperation, focused on the post-ESDP process, as currently represented by the publication of the *Territorial State and Perspectives* and *Territorial Agenda* reports and the other, led by DG Regio, and concerned with the implementation of 'territorial cohesion', primarily via the operation of the Community Strategic Guidelines for the Structural Funds. The key spatial issues of accessibility and mobility continue to coalesce around cohesion policy, especially the form of cohesion linked so closely with European competitiveness by the Third Report on Cohesion.

3. Absence: Environmental Justice

The above analysis of the elements of the ESDP highlights the dominance of the economic agenda at work in the EU's developing spatial planning strategy, both in terms of the values expressed and the use of economic methodologies, as illustrated clearly by the adoption of the concept of 'territorial capital'. In discussing these elements I have tended to draw upon the work of critical geographers, whose discipline encourages them to uncover the 'will to order space' and 'spatial difference' through analysis of maps, images, and info-graphics, as well as texts, and policy discourse. In this part, I move to more legal terrain by critiquing the relative absence of environmental justice issues in the ESDP and later spatial planning initiatives. The ESDP, for example, devotes less than one page to social segregation, poverty, social exclusion, and ghettoization in cities, and fails to consider the distribution of harmful environmental effects such as pollution in certain areas or regions.[168] By focusing on the environmental justice content of the ESDP I have adopted a deliberately narrow perspective. Social exclusion, for example, tends to be the preserve of other areas, even though it has a territorial aspect.[169] However, I would still expect more than a notional recognition of the equality and justice issues involved in planning and development from a European perspective.

Admittedly, some advances have been made in post-ESDP thinking on this issue. ESPON, for example, recognizes equity and justice as issues for discussion in its 2007 *Scenarios* report. This links the idea of a cohesive, balanced, 'reasonably "fair" distribution of wealth and development across Europe's regions'[170] with the European Social Model, interpreted here as the 'negotiated organization of work and distribution of productivity gains, as well as a transfer of some wealth

---

[168] ESDP, p. 66.
[169] See Armstrong's work on social exclusion and the OMC, e.g. 'Tacking Social Exclusion through OMC: Reshaping the Boundaries of European Governance', in Borzel, T. and R. Chichowski (eds.) *The State of the European Union, Vol. 6: Law, Politics and Society* (OUP, 2003).
[170] ESPON, *Scenarios on the Territorial Future of Europe*, p. 56.

from the richest to the poorest regions'.[171] The report also mentions the socio-economic segregation of communities within cities, and proposes that this should be moderated by the targeted intervention of public actors in housing markets, hinting at a market-correcting policy. But there is still little which acknowledges how adverse environmental impacts might affect such segregated communities, beyond a recommendation that risk assessments of local and regional climate hazards should be carried out to target investment in adaptation measures (for example water saving techniques in Southern Europe and flood prevention measures in Northern Europe).[172]

Of course it may be argued that a spatial justice dimension is implicit in the cohesion elements of the ESDP and post-ESDP initiatives. According to the Commission, '[C]ohesion policy pursues the principle that people should not be disadvantaged by where they happen to live and work in the Union',[173] an idea referred to in post-ESDP projects as territorial solidarity. However, as discussed above, territorial solidarity does not amount to a redistributive agenda. Instead, the concept describes the idea of equal opportunities to participate in economic and entrepreneurial activity. Giannakourou sees this transformation or weakening of the traditional conceptual and institutional standards of state spatial welfarism as wholly predictable, given that the European spatial planning perspective has been developed within the constraints of a market-oriented integration process and pursued according to 'a soft and voluntary intergovernmental spatial cooperation'[174] mode of governance. She also rejects a nostalgic call for the lost virtues of state interventionism, arguing instead that the abandonment of hierarchical and centralist spatial action could be seen as a search for new forms of collective spatial action working beyond the state,[175] and possibly between actors at different levels.

To take task with this argument, the loss of the redistributive aspect of state welfarism in the European version of spatial planning or territorial cohesion takes on a different meaning when considered from an environmental justice perspective. Here, I am concerned not so much with the fair distribution of 'goods' (perceived by the *Territorial Agenda* in terms of accessibility and mobility, in sum the products of 'sustainable economic growth'), but rather with the distribution of the related 'bads'—motorways, bridges (remarkably controversial in some quarters, for example the Oresund region),[176] high speed railways, and networks of regional airports, the development and use of which tend to disproportionately affect poor and ethnic minority communities. At EU level, the replacement of the redistributive element of national spatial planning by a form of competitive

---

[171] Ibid.
[172] Ibid., p. 61.
[173] CEC, *Third Cohesion Report*, p. 27.
[174] Giannakourou, supra, 627.
[175] Ibid.
[176] Jensen and Richardson, supra, pp. 189–200.

spatial fairness, with its associated compromises and constraints, may be the inevitable product of inter-governmental activity in this area, as Giannakourou argues so persuasively. However, to accept this replacement uncritically is to ignore the EU's apparent misreading of sustainable development, particularly the connections which can, and should, be made between intergenerational justice (responsibility towards future generations) and distributive justice in the present generation in relation to environmental degradation.

In trying to work out the nature of the relationship between sustainable development and justice, Barry focuses on the desirability of democratizing definitions of 'social development' and 'progress'.[177] He first describes that environmentalists tend not to be 'against "development" and the improvement of the human condition, but rather against certain forms of industrial, undifferentiated economic growth and "modernization" which is equated with "progress"'.[178] For radical environmentalists, the central task involved in reconceptualizing this idea of progress towards a more social form is to democratize 'the capacity and power to choose one's development, and to socially define what development is'.[179] In other words, 'for development and preservation to be integrated, the process of defining development itself and not just its implementation, needs to be democratized'.[180] For Barry the importance of analysing our understanding of progress is to grasp *what* it is we owe future generations, according to the concept of sustainable development, as well as identifying our responsibility to current generations. However, as he and others have also made clear, the responsibility for environmental damage and destruction which is held as constituting an injustice to future generations should not be evenly attributed, and it is here that the issue of environmental and social justice for current generations becomes pertinent:

> ...there is a very real danger that underlying questions of power and inequality will be lost, giving the impression that the poor as much as the rich are equally responsible for the environmental crisis and therefore should take the responsibility for caring for future generations. In short, the debate about intergenerational justice will remain politically impotent unless the intimate connection between it and distributive justice within the present generation in relation to environmental degradation is established.... Ultimately, the best way to ensure justice for future generations is to create the conditions of social justice for the present generation, a central aspect of which is the democratic determination of progress and development.[181]

Here, Barry is concerned primarily with the need to secure *intra*generational justice, as between the developed and developing world, but his argument works equally well when applied to a range of spatial scales, so that environmental and

---

[177] Barry, J., 'Green Politics and Intergenerational Justice: Posterity, Progress and the Environment', in Fairweather, B. et al. (eds.) *Environmental Futures* (Avebury, 1999).
[178] Ibid.   [179] Ibid.
[180] Ibid.   [181] Ibid., p. 70.

social justice should also prevail within local communities, cities, regions, and 'super-regions'. On the practicalities involved in this, Baden et al. have thoroughly examined the difficulties involved in establishing the scale and scope of environmental justice studies in the United States, finding that such studies can range over individual firms, cities, counties, states, sub- and multi-state regions, even the nation as a whole.[182] As an example of the challenges involved, they cite President Clinton's Executive Order 12898 (Federal Actions to Address Environmental Justice in Minority Populations and Low-Income Populations) which required federal agencies to 'make achieving environmental justice part of its mission' and identified minority communities as those where the percentage minority is 'meaningfully greater than the minority population percentage in the general population or other appropriate unit of geographic analysis'.[183] Baden et al. find that the choice of scale and scope can influence the results of analysis, with studies at smaller scales appearing to exhibit more statistically insignificant findings than at larger scales.[184] Overall, however, their analysis of the many studies conducted on environmental justice on various scales and with differing areal units is consistent with early environmental justice studies which show that 'hazards are more likely to be found where minorities or low-income people are'.[185]

The lesson for the EU is that the methodologies for establishing the geographic domain for an environmental justice study might be far from straightforward and there are some variations in result depending on the spatial scale chosen, but that investigations into the state of distribution of environmental risks, hazards, or harms can be undertaken meaningfully at a number of levels. In very strong terms, Baden et al. assert that 'environmental inequity at *any* scale is evidence of injustice and motivates policy to correct the injustice ... just because evidence of injustice is apparent at one scale and not at another does not negate the evidence of inequity'.[186]

Returning to the themes of Barry's theoretical examination of the connections between sustainable development and environmental justice, empirical analysis at local and regional levels is currently commonplace in the United States. Portney, for example, produces a measured assessment of this nexus in the case of waste disposal or treatment facilities, concluding that the nature of the relationship

---

[182] On the application of theories of environmental justice to varying geographical scales, see Baden, B. M. et al., 'Scales of Justice: Is there a Geographic Bias in Environmental Equity Analysis?', *Journal of Environmental Planning and Management* 50/2 (2007) 163.

[183] Council on Environmental Quality, *Environmental Justice: Guidance under the National Environmental Policy Act* (Executive Office of the President, 1997), p. 25, cited by Baden et al., supra, 181.

[184] Baden et al. give the following example : if 'downtowns' host more poor minorities and also more pollution relative to other areas, then restricting the scope of the study to downtowns may undertstate the larger spatial correlation between poor minorities and pollution ('Scales of Justice', 166).

[185] Ibid., 178.

[186] Ibid., 164.

depends in part on which definition of sustainability one uses and on the specific type of action pursued as part of trying to achieve environmental justice.[187] For example, if a waste disposal siting fails to take place because of successful local (LULU/NIMBY (not in my back yard)) opposition, what might be the alternative? Illegal dumping? Or a thorough search to reduce the production of waste in the first place? For Portney, knowing what these alternatives are plays an important part in understanding the nexus between different models of sustainable development (sustainable growth, ecological sustainability, and sustainable communities) and environmental justice.[188]

Work such as this forms part of a more general and active environmental justice movement in the United States, the evolution of which is a familiar success story, involving grassroots activists, academics, and politicians. The origins of the movement are widely attributed to the work of religious and anti-racism groups which revealed a pattern of siting hazardous waste dumps nearest to communities which were predominantly made up of the poor and ethnic minorities, or 'the line of least resistance'.[189] The interpretation of this correlation as a form of environmental racism has since been subject to a great deal of analysis, and this charge is now more subtly understood, particularly in terms of how the dynamics of the real estate market tend to render any attempt to distribute LULUs 'fairly' a futile exercise.[190] The political, legislative, and academic high profile of the environmental justice movement in the United States stands as a marked contrast to the currently weaker position of such concerns in the EU.[191] Most obvious is the overt connection made between 'race and space' in the United States. This helped identify common ground between the civil rights movement with its history of challenging segregation and discrimination and an emerging environmental agenda, producing grassroots resistance to the uneven distribution of hazardous land uses, with some success.[192] Some of the ideals of the environmental justice movement were advanced in law and policy through the use of

---

[187] Portney, 'Environmental Justice and Sustainability: Is There a Critical Nexus in the Case of Waste Disposal or Treatment Facility Siting?' (1993–4) 21 *Fordham Urb L J*, 827.

[188] Ibid., 837.

[189] United Church of Christ, *Commission for Racial Justice* (1987), and update, Bullard, R., Mohai, P., Saha, R., and Wright, B., *Toxic Wastes and Race at Twenty 1987–2007* (United Church of Christ, 2007) <http://www.ucc.org/>.

[190] Been, 'What's Fairness Got To Do With It? Environmental Justice and the Siting of Locally Undesirable Land Uses' (1993) 78 *Cornell L Rev* 1001.

[191] For a comparative approach, see further Layard, 'Environmental Justice: the American Experience and its Possible Application in the United Kingdom', in Holder, J. and McGillivray, D. (eds.) *Locality and Identity: Environmental Issues in Law and Society* (Dartmouth, 1999).

[192] See, for example, Yang, 'Melding Civil Rights and Environmentalism: Finding Environmental Justice's Place in Environmental Regulation' (2002) 26 *Harv Envtl L Rev*. 1; Foster, 'Justice From the Ground Up: Distributive Inequities, Grassroots Resistance, and the Transformative Politics of the Environmental Justice Movement' (1998) 86 *Cal L Rev* 755, and Lazarus, 'Pursuing "Environmental Justice": the Distributional Effects of Environmental Pollution' (1992–3) 87 *NW U L Rev* 787. A critical appraisal of the environmental justice movement is offered by Pellow, D. and Brulle, R. (eds.), *Power, Justice and the Environment* (MIT, 2005).

forms of impact assessment, for example Executive Order 12898, which has led to developers being pressured into changing their siting plans (though apparently the Environmental Protection Agency has fallen short of overturning a state agency decision under federal licence on the basis of a breach of its environmental justice criteria).[193]

This use of environmental assessment offers some insights into how environmental justice might develop in a spatial planning context in the EU, particularly since the building blocks for such an approach already exist in the form of the EIA and SEA Directives.[194] The potential expansion of environmental assessment to take into account whether negative environmental effects are borne disproportionately by certain sectors of the population has been explored, but remains at a formative stage.[195] There are also some serious doubts about the ability of environmental assessment to draw the public in to decision-making about 'fairness' (as with other concerns) in any meaningful way because of the technical and scientific nature of the procedure which works to exclude dialogue with the public, with the result that aesthetic and ethical problems are not taken into account.[196] The existing practice of environmental assessment is aptly described by Elling as reflecting 'a power struggle between production, distribution and the implementation of knowledge in environmental politics' in which 'the winners in this struggle have until now been the planners or administrators'.[197] Jensen and Richardson's research on the (overdue) strategic environmental assessment of the TENT-T supports this analysis. They found that the SEA contains little apparent potential for public participation, rather the assessment model 'appears to resemble an analytical tool to be used in closed policy analysis and decision support rather than to facilitate a participative planning process for TEN-T'.[198]

A key concern in this context is the adequacy or otherwise of public involvement in the development of the European spatial strategy, particularly those aspects set by the TEN-T. Confident claims were made by the CSD about the participatory and transparent nature of the process culminating in the ESDP, but there is also criticism that the consultation process was more exclusive, successfully engaging professional, political, and corporate elites, whilst

---

[193] Stallworthy, M., *Sustainability, Land Use and Environment: A Legal Analysis* (Cavendish, 2002), citing Lyle, 'Reactions to EPA's Interim Guidance: the Growing Battle for Control Over Environmental Justice Decision-Making' (2000) 75 *Indiana LJ* 687.

[194] Directive 85/337/EEC OJ 1985 L 175, p. 40, as amended by Directive 97/11/EC OJ 1997 L 73, p. 5 and Directive 2001/42/EC OJ 2001 L 197, p. 30.

[195] Friends of the Earth, 'Environmental Justice Impact Statement: An Evaluation of Requirements and Tools for Distributional Analysis' (2005) <http://www.foe.co.uk/resource/reports/ej_impact_asessment.pdf>; see also Lichfield, N., *Community Impact Evaluation: Principles and Practice* (UCL Press, 1996).

[196] Elling, B., *Rationality and the Environment: Decision Making in Environmental Politics and Assessment* (Earthscan, 2008).

[197] Ibid., p. 4.

[198] Jensen and Richardson, supra, p. 169.

overlooking the views of local communities in both peripheral and more urbanized regions.[199] Even ESPON has recalled that the ESDP was developed by a limited number of actors',[200] which suggests that the European spatial strategy making process could usefully learn from the collaborative planning movement, which has its roots in the United States but has also transplanted successfully in many of the EU's Member States.[201]

To recap, in the EU the need to develop integrated policy-making as an important element of sustainable development has been appreciated; the ESDP, as with other instruments, represents a genuine attempt to bring this about. But the justice element of sustainable development is lacking in this and later spatial initiatives, particularly when compared to the position in the United States in which the recognition of spatial differences in treatment formed part of a radical agenda for civil rights. Looking beyond the spatial planning agenda, Schwarte and Adebowale argue that more generally within the EU, environmental protection and the fight against racial and ethnic discrimination are rarely considered within an integrated framework.[202] They compare the existence of well established procedural requirements for environmental justice under EU law (implementing the access to environmental justice provisions of the Aarhus Convention)[203] with the current absence from the EU policy agenda of the distributive aspects of environmental justice,[204] and advise using anti-discrimination legislation to challenge decisions and activities which create negative environmental impacts and neglect the interests of Black, Asian, and Minority Ethnic Groups.[205]

4. Conclusions

Sustainable development and environmental justice represent two vitally important environmental 'movements' which have developed rapidly over the past twenty years or so, both in parallel and in tandem,[206] depending on the relative strength of each concept at a particular time. However, as Barry states, it is now clear that the realization of sustainable development (as well as other environmental goals) is firmly underpinned by considerations of justice which

---

[199] Williams, 'Research Networking and Expert Participation in EU Policy Making', AESOP XIII Congress, 7–11 July 1999.
[200] ESPON, *Application and Effects of the ESDP in the Member States*, p. 25.
[201] See, for example, Elster, J. (ed.) *Deliberative Democracy* (CUP, 1998) and Healey, supra.
[202] Schwarte and Adebowale, supra, p. 4.
[203] Convention on Access to Information, Public Participation in Decision-Making and Access to Justice in Environmental Matters (Aarhus Convention) (1995) 38 ILM 515.
[204] Schwarte and Adebowale, *Environmental Justice and Race Equality in the European Union*, p. 4. See also Adebowale, M., *Using the Law: Barriers and Opportunities for Environmental Justice* (Capacity Global, 2007).
[205] Adebowale, M., *BAMEs Tackling Social and Environmental Justice* (Capacity Global, 2005) <http://www.capacity.org.uk/resourcecentre/reports.html>.
[206] Portney, supra, 827.

have international, domestic, as well as intergenerational dimensions. This means that a theory of distributive environmental justice should be regarded as the most ambitious conception of justice to date.[207] In building a spatial dimension into EU policy-making, by the incorporation of European spatial principles in regional funding decisions and more specifically but indirectly by trying to influence the thinking (or 'spatial positioning') of planners and policy-makers in the Member States through the ESDP and more recent initiatives, the central but difficult issue of justice has been sidestepped, or conceived of narrowly, as a matter of territorial cohesion (née spatial planning) which remains closely allied to competitiveness concerns. As a result, this area is now shaped by 'territorial integration', 'territorial capital', 'territorial solidarity', 'territorial governance', and the unifying concept of the 'Territory of the EU', each directed towards enhancing the EU's global position, but there is still no fully worked through conception of territorial justice which might plausibly begin to merge environmental and social justice concerns, to create a Green-Red alliance,[208] in such a framework.

The formation of a spatial planning/territorial cohesion strategy by the EU marks a relatively recent and deliberate shift from specific environmental protection and other concerns to more broad-based land and spatial issues. (This is in clear contrast to the experience in the United Kingdom in which environmental policy emerged from a collection of laws relating to land use and development, notably planning, and the law of nuisance).[209] In the EU, environmental law and policy has furthered the integration of apparently discrete areas, in line with the process of policy mainstreaming, or accommodating an 'environmental counter discourse' within 'business as usual' policy-making networks and structures. Bringing in a spatial dimension offered the opportunity to further integrate decision-making having impacts upon land use patterns and landscapes, but the issues involved and the relationships between the spatial elements are complex and layered, creating also the potential loss of a clear reference point for environmental protection and environmental justice. The main point here is that whilst it is well recognized that social inequality and spatial inequality reinforce each other, these have not been fully integrated in the context of the EU's vision of spatial planning, beyond the 'solidarity' conditions for equality of access to services and equal participation in entrepreneurial activity. To return to the reading of EU spatial policy as a monotopia, *economic* integration remains the governing discourse. The EU's policy drive is to order and overcome space to ensure frictionless mobility of people, goods, services, and capital, in the main through the construction of hard, 'consuming', infrastructure. This constitutes a modern European 'enlightenment' project, with its own rationality of (European)

---

[207] Barry, supra, pp. 57–8.
[208] On the prospects for such an alliance, see Dobson, A., *Justice and the Environment: Concepts of Environmental Sustainability and Dimensions of Social Justice* (OUP, 1998).
[209] Bell, S. and McGillivray, D., *Environmental Law*, (OUP, 2007) Ch. 1.

progress—the legal, technical, and physical achievement of the 'four freedoms'—suggesting that any attempt to integrate the alternative rationalities of environmental and social justice will be partial at best. Considering further the necessary linkage between sustainable development and environmental justice, decisions about the direction of development and the nature of 'progress' as defined by the EU's spatial dimension have not been democratized. This raises important questions about the ability of 'new governance' methods, such as the setting up of expert discursive networks by the ESDP, to fully reflect social and environmental justice concerns. To sum up, even without (until recently) a formal legal base for spatial policy, one sees a growing awareness of 'a common European territory',[210] and the development of a specific European way of thinking about space, rather less so any clear sense of a policy concern or legal protection for the 'losers' caught up in this process.

[210] Jensen and Richardson, supra, p. 210.

5

EC External Relations on Environmental Law

Massimiliano Montini

1. Introduction

The European Community is now an established key player in international relations, despite the relatively weak and cumbersome provisions contained in the EC Treaty in respect thereof. Within the field of the EC Treaty's application, some specific norms upon which EC external policy action is based do in fact exist, but they are very limited in scope and reach. In the environmental sector, in particular, the European Community enjoys an explicit competence to act not only at the internal, but also at the external level. However, this competence is concurrent with that of the Member States and must be exercised in conformity with the relevant provisions of international law, which may sometimes severely limit its full effect.

The present chapter is dedicated to the analysis of EC External Relations in the environmental sector and is divided into three parts.

The first provides a general analysis of EC external relations; here the competence established by the EC Treaty for the European institutions is examined, as well as the contribution of both the case law and the most contemporary legal literature. This is done in order to define the present state of the art in the field of EC external relations with regard to the distribution of powers between the EC institutions and the Member States.

The analysis then proceeds to the specific case of EC external relations in respect of the environment. In particular, the specificities of the sector are scrutinized in order to determine whether and to what extent they differ from the general case of EC external relations.

In the second and third parts of the contribution, two case studies are presented. The first deals with the demarcation of competences between the EC institutions and the Member States in the 'trade and environment' area. In this respect, the specific provisions of the EC Treaty devoted to the EC's common commercial policy and to environmental protection will be compared, followed by an analysis of the most relevant opinion of the European Court of Justice, namely the so-called WTO Opinion (1994) and Cartagena Protocol Opinion (2000).

In the second case study, the analysis shifts to the subject of climate change. The interplay between the EC's external dimension relating to the relevant legal provisions in the field of climate change, represented mostly by the Kyoto Protocol on Climate Change (1997) and subsequent implementing provisions, is analysed and compared with the EC's internal dimension, which refers first of all to the 1998 EC Burden Sharing Agreement, and which also includes all the provisions falling within the realm of the so-called EC Climate Change Policy, composed of several non-binding and binding legal acts.

Finally, the contribution concludes with some brief suggestions that the EC should endeavour to propose/impose its environmental standards and rules to third countries, without violating the applicable international legal provisions, in the framework of a new approach which could be called the 'environmental conditionality' approach. On the basis of such an approach, in fact, the EC could try to promote ways to 'push' other States towards a (more or less) voluntary adoption of EC environmental standards and rules as a reference point for the development of their own national legislation, possibly linking their adoption and effective implementation to the provision of various other advantages for the States concerned in their relations with the EC itself.

2. EC External Relations and the Protection of the Environment

A. Introduction on the Existence and the Exclusivity Question

The matter of EC external relations is traditionally one of the most complex of all EC law. The main reason for this lies in the fact that the EC institutions, at least initially, were not meant to have extensive powers in the external relations field and almost all EC competences and policies were planned and inserted into the EC Treaty with a focus solely on the internal dimension, the only notable exception being EC commercial policy, which by definition ought to possess both an internal and external dimension.

Therefore, the analysis must start from the principle of the attribution of powers, which is contained in Article 5 of the EC Treaty, according to which 'The Community shall act within the limits of the powers conferred upon it by this Treaty and of the objectives assigned to it therein.'

Leaving aside for the moment the basic distinction contained in the EC Treaty between exclusive and concurrent competence of the EC institutions vis-à-vis the Member States, we should concentrate on the basic pillar upon which the

EC Treaty system is based. According to this pillar, no competence may be claimed under EC law by the EC institutions if no explicit reference to this effect exists in the text of the EC Treaty. This is the essential meaning of the principle of the attribution of powers, from which no derogation is foreseen in the Treaty.

Besides this principle, the EC Treaty does not contain any reference to the relationship between the internal and external dimensions of the powers conferred to the EC institutions. It is normally understood from the wording of the principle of the attribution of powers that this is meant to refer only to the internal dimension. Therefore, the issue of the possible extension of EC competence to the external field must be framed in other terms and addressed through other means, as there is no solution provided by the EC Treaty itself.

In this regard, it must be said that under EC law, there is an unwritten principle inherent in the interpretation of the provisions of Article 5 of the Treaty on the attribution of powers, according to which no parallelism exists under EC law between internal and external competences. This means, in other words, that since the EC Treaty normally makes no reference to the possible external competence of the EC institutions pursuant to the Treaty, it cannot be normally inferred by way of interpretation that the EC institutions enjoy external competence in a certain field in all cases where an internal competence in such a sector is provided for by the Treaty. This principle has been named the principle of the 'lack of parallelism between EC internal and external competence'.

However, the application of the above principle is not a completely strict one. In fact, the principle, though never repudiated but confirmed by the European Court of Justice which stated that 'the Community [...] has only those powers which have been conferred upon it',[1] has been tempered through the years by the ECJ case law. The Court has, in fact, decided on several controversies by identifying a series of cases in which the existence of an EC external competence in a given situation—under very precise and strict conditions—may be inferred by an explicit internal competence conferred upon the EC institutions by the Treaty.

The various, complex and not always very consistent ECJ case law relevant to this matter has been grouped in the legal literature under two headings, which in turn correspond to the two very basic questions of principle: the existence question and the exclusivity question.[2]

B. The Existence Question

The existence question departs from the premise that the provisions of the EC Treaty devoted to external relations and the corresponding external competences

---

[1] See Opinion 2/94, [1996] ECR I-1759, § 23.
[2] The distinction between the existence question and the exclusivity question is taken from A. Dashwood and was originally contained in Dashwood, A., 'Implied External Competence of the EC', in Koskenniemi, M. (ed.), *International Law Aspects of the European Union* (Kluwer Law International, 1998), p. 113.

are minimal and incomplete.³ As a consequence, there are several actions and initiatives which are to be undertaken by the EC institutions at the international level in order to fulfil the EC Treaty's main objectives, which do not find a solid and clear legal basis within the Treaty itself. For this reason, the European Court of Justice has made use of its well-known capability to fill the gaps in the legal order established by the EC Treaty by developing a specific doctrine on the 'implied external relations competence' of the European Community. A reference statement for this doctrine may be the following, contained in ECJ Opinion 2/94 relating to the possible accession of the EC to the European Convention on Human Rights, according to which 'The Community acts ordinarily on the basis of specific powers which [...] are not necessarily the express consequence of specific provisions of the Treaty but may also be implied from them.'⁴

The above statement, in fact, calls for an extensive interpretation of the principle of the attribution of powers as contained in Article 5 EC Treaty, according to which the powers conferred to the Community by the Treaty may be interpreted to refer not only to the powers explicitly conferred by the EC Treaty to the EC institutions, but also those powers which may be implied, by way of interpretation, from the provisions of the EC Treaty itself. Going back to the analysis of the ECJ in this respect, it emerges that in the Court's view:

> whenever Community law has created for the institutions of the Community powers within its internal system for the purpose of attaining a specific objective, the Community is empowered to enter into the international commitments necessary for attainment of that objective even in the absence of an express provision to that effect.⁵

In sum, ECJ case law on the existence question may be summarized in three main principles, namely the attribution of powers principle, the implied external powers principle, and the effectiveness principle.⁶

The principle of the attribution of powers states what is immanent in the EC legal order, that is, the fact that the European Community has a normative competence only in those sectors and the extent to which it has been explicitly conferred with powers by the EC Treaty or other relevant sources of EC law. Moreover, EC competence is limited by the principles of subsidiarity and proportionality, which, pursuant to Article 5 EC Treaty, integrate the principle of the attribution of powers in order to protect the prerogatives of the Member States once Community competence has been ascertained.

---

³ On the existence question see Dashwood, A., 'The Attribution of External Relations Competence', in Dashwood, A. and Hillion, C. (eds.), *The General Law of EC External Relations* (Sweet & Maxwell, 2000), p. 115ff.
⁴ See Opinion 2/94, supra n. 1, § 25 (recalling previous ECJ case law on the matter).
⁵ Ibid. § 26 (recalling previous ECJ case law on the matter).
⁶ See Montini, M., 'The EC's External Competence and the Protection of the Environment', in Jans, J. H., (ed.) *The European Convention and the Future of European Environmental Law* (Europa Law Publishing, 2003), p. 70 at 73.

The implied external powers principle, then, contributes to shaping more precisely the principle of the attribution of powers in the sense that the Community must not always base its external competence on explicit EC Treaty provisions (although this is the normal situation). In some cases, such a competence may be implicitly inferred from other provisions of the Treaty, pursuant to Article 308 EC Treaty, as clearly emerges from the already mentioned Opinion 2/94.[7]

Much earlier than in the above Opinion, however, the origin of the implied external powers principle can be traced back to the *AETR* case.[8] Yet here, the reasoning of the Court did not determine the reach of the implied external powers principle in a very clear way. Indeed, even if it could be inferred that while the possibility of deducing the existence of a Community external competence from the existence of an internal competence is implicitly admitted, such a possibility seems to be limited to situations where the said internal competence has been concretely exercised. That is to say, the decision of the ECJ in the *AETR* case leaves the question open as to whether an implied EC external competence may arise in a situation where the corresponding internal competence has not yet been exercised.

This question was in fact better addressed by the ECJ in the subsequent 1976 *Kramer* case,[9] pertaining to the distribution of competence in the field of the conservation of marine resources. In this case, the Court affirmed that despite the fact that the Community had no internal legislation in place in the field of the conservation of marine resources, it nevertheless had the power to conclude international agreements for the conservation of marine resources. In its subsequent case law, the ECJ reiterated the implied external powers principle in various situations, such as in the already cited Opinion 1/76 and 2/94.[10] Moreover, the reach of such a principle was reinforced by the emergence of the effectiveness principle.

The effectiveness principle, also known as the 'necessary attainment principle', was first coined by the ECJ in Opinion 1/76,[11] which dealt with the question of the boundaries of EC competence to conclude an international agreement establishing a European fund for inland waterway vessels. The principle essentially determines that whenever Community law has created certain specific powers for the EC institutions with a view to attaining a specific objective and whenever 'the participation of the Community in the international agreement is necessary for the attainment of one the objectives of the Community',[12] the

---

[7] Ibid. § 25–6.
[8] See Dashwood, 'Implied External Competence of the EC', supra n. 2, at 117–18, which traces back the origin of the implied external powers principle to the *AETR* case, infra n. 16.
[9] See Joined Cases 3, 4, 6/76 (*Kramer* case), [1976] ECR 1279.
[10] See for instance Opinion 1/76, [1977] ECR 741, § 3, and Opinion 2/94, supra n. 1, § 25–6.
[11] See Opinion 1/76, supra n. 10.
[12] Ibid. § 3–4.

EC can enter into the envisaged international commitments despite the absence of an explicit provision to this effect contained in the text of the EC Treaty, even if the corresponding internal power has not yet been exercised.

Thus the effectiveness principle has been expressed in similar terms in various other ECJ cases, among which a paramount role is played once again by Opinion 2/94.[13] In the said Opinion, the Court recalled in very clear terms that whenever the EC Treaty confers specific internal powers on the EC institutions in order to attain a certain objective, 'the Community is empowered to enter into the international commitments necessary for attainment of that objective even in the absence of an express provision to that effect'.[14] This means, in other words, that when the achievement of a certain objective set out in the EC Treaty could be endangered where the Community acts internally, but not externally, it seems to be more effective to recognize specific external powers of the EC institutions in the sector(s) concerned, even if the EC Treaty does not explicitly provide to this effect.

C. The Exclusivity Question

Once the competence of the European Community in a given sector has been ascertained, the exclusivity question may then come into play.[15] This question departs from the consideration that once the European Community has exercised its external competence and has laid down common rules in a given sector, Member States should still enjoy an unlimited right to carry out obligations with third countries in the external relations field. The outcome might be a situation where EC common rules would be negatively affected by the obligations carried out by one or more Member States. As a consequence, according to ECJ case law, it must be held that when there is a risk of negative interference arising from the definition and application of some national provisions stemming from international obligations, Member States should refrain from adopting and implementing them so as not to jeopardize the existing common EC rules regulating the matter.

In other words, in the ECJ's view—firstly expressed in the decisive *AETR* case[16]—in areas of shared or concurrent competence between the EC institutions and the Member States, the actual 'use' of powers by the EC institutions, in case 'common rules' are adopted which fully regulate a certain matter, would have the effect of pre-empting the possibility for EC Member States to act in that

---

[13] See Opinion 2/94, supra n. 1. See also ECJ Case C-476/98 *Commission v. Germany* (*Open Skies* case), [2002] ECR I-9855, § 80ff.
[14] See Opinion 2/94, supra n. 1, § 26.
[15] On the exclusivity question see O'Keeffe, D., 'Exclusive, Concurrent and Shared Competence', in Dashwood, A. and Hillion, C. (eds.), *The General Law of EC External Relations*, (Sweet & Maxwell, 2000), p. 179ff.
[16] See Case 22/70 *Commission v. Council* (*AETR* case), [1971] ECR 263.

particular field with possibly conflicting national provisions. In fact, under the so-called *AETR* doctrine, the Court affirmed that:

each time the Community, with a view to implementing a common policy envisaged by the Treaty, adopts provisions laying down common rules, whatever form these may take, the Member States no longer have the right acting individually or even collectively, to undertake obligations with third countries which affect those rules. As and when such common rules come into being, the Community alone is in a position to assume and carry out contractual obligations towards third countries affecting the whole sphere of application of the Community legal system.[17]

As has been aptly stated in the legal literature, 'the rationale for the doctrine is to avoid unilateral or collective action by Member States which would conflict with internal Community measures'.[18] In fact, the *AETR* doctrine draws a parallel between the 'internal level' prohibition for the Member States to adopt national provisions conflicting with EC ones—once the EC has exercised its powers in a given sector—which derives from the 'principle of supremacy' of EC law over national law, and the 'external level' prohibition for the Member States to enter into international obligations which may negatively affect the common rules that have been adopted by the European Community in a given sector. Therefore, according to the *AETR* doctrine, when the EC has adopted legislation which fully or partially regulates a certain matter, the existence of a Community measure makes EC competence in that field an exclusive one.

In sum, following the interpretation given in the most relevant legal literature, the ECJ's case law on the exclusivity question may be summarized in two main principles, namely the exclusivity principle and the loyalty principle.[19]

The exclusivity principle originates from the ECJ decision delivered in the 1971 *AETR* case, specifically in paragraphs 17 and 22. In particular, paragraph 17 states that the Member States do not have the right, either individually or collectively, to undertake obligations with third countries which may 'affect' the common rules that have already been laid down by the Community with a view to implementing a common policy envisaged in the EC Treaty.[20] Thus, the exclusivity principle creates a parallel between internal competence, when it has been concretely exercised by the Community through the adoption of common rules, and the relative external competence. This reasoning is strengthened by paragraph 22 of the decision, which states:

To the extent to which Community rules are promulgated for the attainment of the objectives of the Treaty, the Member States cannot, outside the framework of the Community institutions, assume obligations which might affect those rules or alter their scope.[21]

[17] See § 17 of the *AETR* case, supra n. 16.
[18] See O'Keeffe, p. 182.
[19] See Montini, pp. 74–5.
[20] See § 17 of the *AETR* case, supra n. 16.
[21] Ibid. § 22.

The exclusivity principle was then restated by the Court in Opinion 1/76 and further refined in Opinion 1/94,[22] where the Court clarified and somehow limited the extent of the principle by stating that: 'The Member States, whether acting individually or collectively, only lose their right to assume obligations with non-member countries as and when common rules which could be affected by those obligations come into being'.[23]

Besides the exclusivity principle, the loyalty principle, as enshrined in Article 10 EC Treaty, is also relevant in this context. It provides for a duty of abstention for Member States from behaviour which may forfeit their obligations under the EC Treaty or secondary Community legislation. That is to say, on the basis of the loyalty principle, Member States should refrain from undertaking international obligations not only in those sectors which fall within the sphere of the Community's common policies established by the EC Treaty, but also in all other possible legislative sectors when there is a risk that the adoption of national provisions in a certain field may affect the existing EC common rules.

However, in order not to overestimate the reach of such a principle, when interpreting the relevant ECJ case law, it is probably necessary to distinguish between those cases in which the EC legislation has reached a complete harmonization and those where it has simply set some minimum standards.[24] In the first, on the basis of the ECJ case law it may be held that the European Community enjoys exclusive competence in the given field both at the internal and the external level. Moreover, in such a case the adoption of specific EC legislation which has achieved complete harmonization on the internal level pre-empts the adoption of either internal or external measures by the Member States.[25] In the second case, namely the one where the EC has simply set some minimum standards, Member States still have the right, in principle, to undertake international obligations and conclude international agreements which set higher standards. In such a case, in fact, the EC and its Member States maintain a concurrent external competence.[26]

D. EC External Relations in the Environmental Field

Now that the issue of EC external relations has been analysed from a general perspective, it is time to deepen the present analysis with reference to the specific field of environmental protection, having regard first of all to the relevant EC Treaty provision and second to the relevant ECJ case law on the matter.[27]

---

[22] See Opinion 1/94 [1994] ECR 1–5267.   [23] Ibid. § 77.
[24] See O'Keeffe, supra n. 15.
[25] See for instance § 17–18 of the *AETR* case, supra n. 16, and § 96 of Opinion 1/94, supra n. 22.
[26] See for instance § 18–21 of Opinion 2/91 (Opinion on ILO Convention) [1993] ECR I-1061.
[27] Please note that in the present contribution, a reference is made generally to the EC Treaty currently in force, normally not taking into account the amendments agreed with the Lisbon Treaty, signed in December 2007 and due to enter into force in January 2009, provided that its ratification is concluded on time by all the 27 EC Member States.

The legal foundation for the exercise of EC environmental policy and law is embodied in Article 174 EC Treaty, which contains a reference to the objectives, relevant principles, and conditions for European legal action in this field.[28] As far as the EC's external competence to act in the area of environmental protection is concerned, there is a specific paragraph in the article which deserves our analysis, namely Article 174(4), which reads as follows:

Within their respective spheres of competence, the Community and the Member States shall cooperate with third countries and with the competent international organisations. The arrangements for Community cooperation may be the subject of agreements between the Community and the third parties concerned, which shall be negotiated and concluded in accordance with Article 300.

The previous subparagraph shall be without prejudice to Member States' competence to negotiate in international bodies and to conclude international agreements.

The above mentioned provision re-states the fact that both the EC and its Member States may become Parties to international agreements in the environmental sector. This is, however, not surprising insofar as the environmental sector represents an area of 'concurrent competence' between the EC and its Member States. This means that, in principle, both can act either at the internal or at the international level and can adopt the necessary legislative and administrative provisions to give effect to the policy choices made.

In fact, the provision mentioned above, namely Article 174(4) EC Treaty, simply affirms in an explicit way the EC's external competence in the environmental field, which, on the basis of the ECJ doctrine described above, could have been inferred anyway from the existence of internal EC competence in the same field—with the aid of the interpretative principle seen above and acting within the framework of the so-called existence and exclusivity questions.

As far as the concrete ways to exercise its external competences in the environmental field are concerned, it must be noted that the provision at hand makes an explicit reference to the fact that international agreements entered into by the European Community must be negotiated under the provisions of Article 300 EC Treaty.[29] This is what is relevant from the EC internal standpoint. Needless to say, there is also an international perspective which matters for the exercise of EC external power in this field, consisting in the prerequisite that the international environmental treaty to which the European Community wishes to

---

[28] On the European external environmental policy in general see Jans, J. H., *European Environmental Law*, 2nd edn (Europa Law Publishing, 2000), p. 69ff.

[29] It should be noted here that an amendment to Article 174(4) EC Treaty contained in the 2007 Lisbon Treaty removes the reference to Article 300 EC Treaty as the provision containing the standard procedure for negotiating an international treaty by the European Community. However, since the process of ratification of the Lisbon Treaty (2007) by the EC Member States is currently still under way, it seems too early to comment extensively on the possible implications of such an amendment.

accede enables an international (economic) organization, such as the EC, to become a Member together with its Member States.

Nowadays, most international environmental agreements foresee the possibility of the EC becoming a fully fledged Party together with its Member States. Some examples of this include the Vienna Convention and Montreal Protocol for the Protection of the Ozone Layer, the UN Framework Convention and the Kyoto Protocol on Climate Change, the Biodiversity Convention and the Biosafety Protocol. The most notable (and probably only) exception is the 1973 CITES Convention on International Trade in Endangered Species, a very early environmental protection treaty which did not foresee any possibility of membership for the EC. This approach, however, was not taken up in more recent environmental treaties, which instead always purport to give full membership to the European Community.

Besides acceding to international environmental treaties, the EC may also act in other ways which could have some relevance for the external relations field. This is the case, for instance, with the adoption of internal legislation which aims at pursuing extraterritorial environmental protection goals. The issue of the legitimacy of internal EC measures with an extraterritorial reach logically entails two different aspects for our analysis; the first is the issue of the legitimacy of their adoption on the basis of the EC Treaty (internal dimension) and the second is the legitimacy of such norms under the pertinent provisions of public international law, namely those relating to the freedom of international trade as guaranteed by WTO rules (external dimension).

As far as the internal dimension is concerned, the starting point for the analysis ought to be Article 299 EC Treaty, which determines the scope of the territorial application of the EC Treaty as well as EC Law altogether. The reference in this provision is made simply to the Member States which are parties to the EC Treaty, implicitly including the territory over which they exercise their jurisdiction, as well as some overseas territories which some Member States consider a fully integrated part of their territory, such as the French Overseas Departments for the French Republic, the Azores and Madeira for the Portuguese Republic, and the Canary Islands for the Kingdom of Spain. Such a broad reference to the scope of application of EC Law implies that there is no limitation to the possibility of the EC institutions adopting legislation which applies to all the territories under the sovereignty of the EC Member States, including the overseas territories mentioned above. However, this does not necessarily give the European Community the power to adopt legislation with an extraterritorial dimension. The possibility to do so, which is recognized under international law for national states, may therefore also be held viable for the European Community with regard to those areas where it enjoys a specific external competence, be it an explicit or an implicit one, only if a different legal basis outside the EC Treaty is found.

In this respect, it seems that such a possibility for the European Community might derive from the exercise of a 'functional jurisdiction', insofar as it is related to the exercise of some of the powers conferred on the European Community by the EC Treaty itself. This issue emerged in the ECJ case law in the *Kramer* case, where the most relevant issue at stake was the extent to which EC Law could regulate the matter of fishing in the high seas, in an area in which the EC clearly enjoyed an internal competence to adopt measures for the conservation of marine biological resources. The ECJ held in such a circumstance that since the EC enjoyed an internal competence: 'it follows [...] from the very nature of things that the rule-making authority of the Community ratione materiae also extends—insofar as Member States have similar authority under public international law—to fishing on the high seas.'[30]

Such an approach was also confirmed in a later ECJ case, namely the *Drift-Nets* case, in which the validity of an EC provision prohibiting the use of drift-nets longer than 2.5 km for the conservation of fishery resources, which had been contested by some fishing companies, was confirmed by the European Court of Justice on the basis of the following reasoning: 'with regard to the high seas, the Community has the same rule-making authority in matters within its jurisdiction as that conferred under international law on the State whose flag the vessel is flying or in which it is registered'. As a consequence, in the material case at stake 'the Community has competence to adopt, for vessels flying the flag of a Member State or registered in a Member State, measures for the conservation of the fishery resources of the high seas'.[31]

What is more relevant for the sake of our analysis is the fact that in both of the cases cited above, the ECJ adopted a similar line of reasoning, which can be described as a 'functional' one with regard to the external relations issues involved. The consequence of such decisions rendered by the European Court of Justice is that it ought to be held that once and to the extent that Member States are competent under international law to adopt measures with an extraterritorial reach, aimed at the protection of the environment outside their territory, the EC should also be considered competent to adopt protection measures in the same field, with the implicit limit that such measures obviously fall under the scope of application of Article 174 EC Treaty. To this effect, however, as has been correctly argued in the legal literature, the provision of Article 174(1), fourth indent, should not be interpreted restrictively, in the sense that the EC's competence to adopt measures with an extraterritorial dimension should not be limited to those areas where 'regional or worldwide environmental problems' are at stake.[32]

With regard to the external dimension mentioned above, that is, the issue of the legitimacy of EC provisions with an extraterritorial scope under the relevant

---

[30] See *Kramer* case, supra n. 9, § 30–3.
[31] See ECJ Case C-405/92 (*Drift-Nets* case), in 1993 [ECR] I-6133, at §§ 12 and 15.
[32] See Jans, p. 72.

provisions of international law, the compatibility of such EC norms with the international rules relating to the freedom of international trade as guaranteed by the WTO rules, most notably those of the GATT, has to be analysed. In order to give proper weight to such an issue, this will be the specific subject of section 3 of the present contribution, which is dedicated to the examination of EC external relations in trade and the environment field.

E. The Existence and the Exclusivity Question in the Environmental Field

After the general analysis on the issue of EC external relations in the environmental field, it is now time to shift the focus to the existence and the exclusivity question, in order to ascertain whether they have some specific and distinct features with respect to the general analysis conducted above.[33]

In this regard, it must be highlighted that first of all, from the environmental protection point of view, since the specific 'Title on the Environment' was introduced in the EC Treaty with the 1986 Single European Act, no problem whatsoever has existed with regard to the existence question. The relevant provision of the EC Treaty, namely Article 174(4) EC, explicitly states that within their respective spheres of competence, both the European Community and the Member States are empowered to conclude agreements with third countries and international organizations. Reality, however, has proved to be a bit more complex than envisaged and controversies among the EC Commission and the EC Council about the correct legal basis for the conclusion of international agreements aimed at the protection of the environment have continued also in recent years. The issue was addressed by the European Court of Justice in Opinion 2/00 on the EC's accession to the Cartagena Protocol on Biosafety.[34] Here, the Court held that since the Protocol had the predominant, albeit non-exclusive, aim of environmental protection, the correct legal basis for its adoption was Article 175(1) EC Treaty, as argued by the EC Council and the Member States, rather than Article 174(4) which was supported by the EC Commission.[35]

As far as the exclusivity question is concerned, it should be underlined that the European Community may, in the environmental field, adopt internal provisions directed either at achieving a complete harmonization or at setting some minimum standards. As a consequence, in the former, Member States are prevented

---

[33] See Montini, pp. 75–6. On this issue see also Chalmers, D., 'External Relations and the Periphery of EU Environmental Law', in Weiss, F., Denters, E., De Waart, P., *International Law with a Human Face* (Kluwer Law International, 1998); Jans, J. H., *European Environmental Law*, (Europa Law Publishing, 2000), p. 69ff.; Thieme, D., 'European Community External Relations in the Field of the Environment' (2001) *European Environmental Law Review* 252ff.; Loibl, G. 'The Role of the European Union in the Formation of International Environmental Law', (2002) *Yearbook of European Environmental Law* 223ff.

[34] On this issue see infra n. 45.

[35] See Krämer, L., *EC Environmental Law*, 5th edn (2003), p. 85.

from exercising any form of external competence, whereas in the latter, the parallel existence of a concurrent EC and Member State external competence is indeed possible. This means that, in a case where a certain matter in the environmental sector has not been completely harmonized by the European Community, the Member States still retain the possibility of fulfilling international obligations independently from the EC institutions in the same field.

However, this leads to the question of whether some limits exist to the conclusion of such external agreements by the Member States. The answer to this question is likely to be found in the wording of Article 174(4) EC, which suggests that no specific limits are imposed by the provision in question regarding the concurrent external competence of the Member States in the environmental field. One can therefore legitimately argue that the only limits effectively posed by the EC legal order on the right of a Member State to act in such a case are therefore embodied by the general limitation inherent in the 'loyalty principle', as enshrined in Article 5 EC Treaty.

The 'loyalty principle', as already mentioned above, states that Member States must refrain from behaving in a way which runs counter to the fulfilment of their obligations under EC law or to the effectiveness of such obligations. However, in the environmental field, a concrete conflict between international obligations undertaken by the European Community and other international obligations independently undertaken by the Member States is not likely to occur. The practice, in fact, shows that the Community and the Member States have usually chosen to act in this field by concluding 'mixed agreements', negotiated by both the EC and the Member States, signed and ratified separately but in a coordinated way, and then jointly executed within the framework of their respective competence.[36]

The conclusion of 'mixed agreements' between the EC and its Member States is in fact a very practical way of entering into international commitments in cases of shared or concurrent competence, such as in the field of environmental law, insofar as it tends to reduce to the minimum the risk of undertaking different and conflicting obligations by the Community and Member States in the same matter.

However, such a practice is not completely free of any shortcomings. The practice of concluding 'mixed agreements', in fact, raises various unresolved questions which can be summarized as follows:

- Where does the effective centre of power lie when 'mixed agreements' in the environmental field are negotiated, adopted, and implemented? Is it held by the European Community or by the Member States? Are 'mixed agreements' the real expression of shared competences between the EC and its Member

---

[36] On the topic see Rosas, A., 'The European Union and mixed agreements', in Dashwood, supra n. 3, p. 200ff.

States or do they become a way for the EC to overwhelm Member States and reduce the scope of their powers?

- Moreover, a problem of 'democracy' in the EC legal order may arise when the conclusion of international agreements pre-empts and/or substitutes the adoption of internal legislation by the EC institutions, and when the procedure for the adoption of international agreement by the Community (contained in Article 300 EC) only foresees a limited competence for the European Parliament and places the 'political' power in the hands of the Commission and the 'legal' power in the hands of the Council.
- When both the EC and its Member States are Parties to a treaty, it is not always completely clear for the other Parties to understand the precise boundaries of their respective spheres of competence. This may lead both to legal problems (which party is responsible in cases of failure to fulfil the obligations stemming from the agreement?) and practical problems (which party must be addressed with regard to questions concerning the implementation of the agreement?). In order to try and solve this problem, the European Community, in cases of 'mixed agreements' normally annexes to its instrument of ratification or accession a 'declaration on the respective EC and Member States competence' which should help to clarify the boundaries of the respective spheres of competence and the corresponding responsibilities with respect to the implementation of the agreements.
- Finally, what about the international responsibility of the Community and the Member States? Broadly speaking, under a mixed agreement both the Community and the Member States concerned might, in theory, be jointly responsible at the international level for the non-fulfilment of the obligations arising from the said mixed agreement.[37] However, the scope of their respective competence, and consequently, their respective responsibilities, may somehow be clarified by the instruments of ratification or accession to an international agreement.[38] Alternatively, their competence and responsibilities under international law may be better defined with the signature of an 'internal agreement' between the EC institutions and the Member States.[39]

---

[37] See for instance Article 4(6) of the Kyoto Protocol on Climate Change (available at <http://www.unfccc.org>), which establishes, in accordance with Article 24(2) of the same Protocol, that the Community and the Member States are jointly responsible for the fulfilment by the Community of its commitments under the Protocol.

[38] See for instance the Declaration by the European Community made in accordance with Article 24(3) of the Kyoto Protocol, which determines the respective spheres of competence for the Community and its Member States in the field of environmental protection. The Declaration is annexed to Council Decision 2002/358 of 25 April 2002 on the approval by the European Community of the Kyoto Protocol, published in [2002] OJ L130/1.

[39] See for instance the Intergovernmental Agreement reached by the EC Member States' Environmental Ministers on the contribution of each Member State to the overall commitment assigned to the European Community by the Kyoto Protocol, Doc. 9702/98 of 19 June 1998 of the Council of the European Union.

In this respect there might be a link between the existence of a relationship between the international responsibility of the Community and its Member States and the responsibility of the Members States towards the Community under EC law. On the basis of this, when a Member State fails to fulfil its obligations under a mixed agreement to which both the Community and the Members States are Parties, in a case where such a violation is a violation of the agreement's obligations by the European Community as well, then the Member State, besides its international responsibility towards the other Parties to the agreement, also bears responsibility towards the Community under EC Law.

The topic related to the potential interplay between the possible issues of international responsibility of both the EC and its Member States in cases of non-fulfilment of obligations stemming from mixed agreements, particularly in the environmental field, and the possible issues of Member States' responsibility towards the European Community under EC Law, will be subject to a more detailed analysis in section 4 of the present contribution, which is dedicated to a case study of the EC's external relations in the climate change sector.

In fact, after this introduction on the subject of EC external relations, which was supplemented by the specific analysis of the EC Treaty provisions relevant for EC External Relations in the field of environmental protection, the two following sections will shift the focus to two specific sectors, namely that of trade and environment and the sector of climate change. In this respect, two case studies will be presented which analyse how EC external relations work in these two key areas and the kind of issues that may arise in the two contexts, depending both on the peculiarities of the specific fields and on the different demarcation of competences between the EC and its Member States in the sectors considered.

3. Case Study on EC External Relations on Trade and the Environment

A. Introduction on the Trade and Environment Issue

The 'trade and environment' sector, in short, refers to all the controversies which may arise when the fundamental freedom of trade comes into contrast with other legitimate interests, such the protection of the environment or the safeguarding of human, animal, or plant health.[40]

---

[40] On the 'trade and environment' issue see Montini, M., 'International Trade and Environmental Protection', in Macrory, R. (ed.), *30 Years of EU Environmental Law: a High Level of Protection?* (Europa Law Publishing, 2006), p. 529ff. On the same topic see also Charnovitz, S., 'Exploring the Environmental Exceptions in GATT Article XX', *Journal of World Trade* 25 (1991) 37; Esty, D., *Greening the GATT: Trade, Environment, and the Future* (Institute for International Economics, 1994); Cameron, J., Demaret, P., and Geradin, P. (eds.), *Trade and the Environment: the Search for a Balance* (Cameron May, 1994) (and in particular the following: Jackson, J., 'Greening of the GATT: Trade Rules and Environmental Policy', 39ff.; Demaret, P., 'TREMS, Multilateralism,

On one hand, this issue may arise at the international level, in which case the EC's external competence in both trade and environment becomes relevant and the EC may be opposed to other World Trade Organization (WTO) Parties, whose right of free trade has been restricted by EC internal measures with an extraterritorial reach aimed at the protection of the environment or public health.

On the other hand, however, the same issue may arise, in very similar terms, at the European Community level, in which case it concerns EC internal competence in the field of trade and environmental or health protection vis-à-vis the Member States. In both cases, similar balancing tools and criteria have been developed and applied by international institutions with adjudicatory functions, such as the Panels and the Appellate Body operating in the framework of the Dispute Settlement Body within the WTO legal order, as well as the European Court of Justice within the European Union legal system.

For the purposes of the present contribution, it is the EC external dimension which matters, and therefore the following analysis will be focused solely on the first of the two dimensions mentioned above, namely that which may oppose the EC to other WTO Parties before the WTO institutions (WTO Panels and the Appellate Body) operating for the resolution of controversies entailing an international trade dimension.

To this effect, the two starting points for our analysis ought to be two specific provisions of the EC Treaty, which are dedicated to EC competence in the 'common commercial policy' area (Article 133 EC Treaty) and the environmental sector (Article 174 EC Treaty) respectively. The analysis of such provisions may help us to determine the legal basis for the EC's participation in the WTO legal system and its boundaries, in order to compare it with the provision on EC external competence in the environmental field.

As far as EC competence in the trade or commercial policy sector is concerned, Article 133(1) EC Treaty states that 'the common commercial policy shall be based on uniform principles, particularly in regard to changes in tariff rates, the

---

Unilateralism and the GATT', in Cameron, J., Demaret, P., and Geradin, P. (eds.), *Trade and the Environment: the Search for a Balance* (Cameron May, 1994), p. 52ff.; Petersmann, E. U., *Trade and Environmental Protection: Practice of GATT and the EC Compared*, (Cameron May, 1994), p. 147ff.); Petersmann, E.U., *International and European Trade and Environmental Law after the Uruguay Round* (Kluwer, 1995); Geradin, D., *Trade and the Environment: A Comparative Study of EC and US Law* (Cambridge University Press, 1997); Montini, M., 'The Nature and Function of the Necessity and Proportionality Principles in the Trade and Environment Context', in *Review of European Community and International Environmental Law* (1997), at 121ff.; Wiers, J., 'Regional and Global Approaches to Trade and Environment: the EC and the WTO', in *Legal Issues of European Integration* (1998), 93ff.; Trebilcock, M. J. and Howse, R., *The Regulation of International Trade*, 2nd edn., (Routledge, 1999); Weiler, J. (ed.), *The EU, the WTO and the NAFTA: Towards a Common Law of International Trade* (Oxford University Press, 2000); French, D., 'The Changing Structure of Environmental Protection: Recent Developments regarding Trade and the Environment in the European Union and the World Trade Organization', (2000) *Netherlands Yearbook of International Law* 1ff.; Cheyne, I., 'Law and Ethics in the Trade and Environment Debate: Tuna, Dolphins and Turtles', (2000) *Journal of Environmental Law* 293ff.; Rao, P. K., *World Trade Organization and the Environment* (Macmillan Press, 2000); Francioni, F., *Environment, Human Rights and International Trade* (Hart Publishing, 2001).

conclusion of tariff and trade agreements, the achievement of uniformity in measures of liberalisation, export policy and measures to protect trade such as those to be taken in the event of dumping or subsidies'.

As one can see, such a provision not only represents a legal basis for EC internal trade or commercial policy competence, but also explicitly establishes the EC's external power to conclude international agreements with third parties in the tariffs and trade sector. It is therefore one of the rare situations within the EC Treaty where both EC internal and external competences are directly established by a Treaty provision itself.

The fact that Article 133 EC Treaty clearly establishes both internal and external competence for the European Community in the common commercial policy sector does not, however, solve all the possible problems relating to the implementation of such a provision. In fact, despite the existing parallel between internal and external powers in such a context, the issue of determining the exact boundaries of EC external policy competence in the trade and commercial sector has in fact arisen and has been the subject of a paramount ECJ interpretative Opinion, namely Opinion 1/94, relating to the participation of the EC in the WTO legal system.

In Opinion 1/94—delivered after the conclusion of the Uruguay Round Multilateral Trade Negotiations and just before the accession of the European Union to the WTO Agreement and all related international trade treaties—the European Court of Justice, in fact, addressed several issues relating to the scope and reach of EC external competence in the trade and commercial policy field. In particular, the ECJ found that the trade in goods, as regulated under the General Agreement on Tariffs and Trade (GATT) fell within the scope of Article 133 EC Treaty, whereas for some of the other WTO related and administered trade agreements the matter deserved more careful scrutiny.

With regard for instance to the General Agreement on Trade in Services (GATS), the Court found that the EC's external competence in that field could not be ascertained once and for all with regard to the whole sector of trade in services, but the analysis had to be articulated along the different modes of supply of services. Following this line of reasoning, the Court found that only the cross-frontier supply of services that do not involve any movement of persons could be assimilated to trade in goods and therefore could be considered to fall within the framework of EC common commercial policy competence, whereas all the other types of supply of services were outside the scope of Article 133 EC Treaty and, as a consequence, fell within the Member States' residual competences. As a result, the ECJ concluded that the EC and its Member States were jointly competent to conclude the GATS Agreement and that such a treaty ought, in fact, to be the object of a 'mixed agreement' concluded both by the EC and its Member States, acting within their respective spheres of competence.[41]

---

[41] It should be recalled here that, in the ECJ Opinion 1/94, the ECJ gave a similar answer to the similar issue regarding the competence to conclude another relevant WTO related treaty, namely the Agreement on the Trade Related Aspects of Intellectual Property Rights (TRIPS).

In the original 1957 EC Treaty, besides Article 133 (formerly Article 113 EC Treaty), there was only one other Treaty provision explicitly empowering the EC to enter into international treaties, namely Article 310 (formerly Article 238 EC Treaty), which deals with the association agreements that may be concluded by the European Community with third countries to pursue common objectives. It was with the 1986 Single European Act that the original EC Treaty was amended and four specific provisions were added. This conferred on the EC institutions a specific competence to enter into international agreements for the purpose of developing cooperation with third parties to better fulfil specific EC Treaty objectives in certain sectors. One of these four sectors was, in fact, the environment, in respect of which Article 174(4) was adopted and which explicitly determines the EC's competence to conclude agreements with third parties in the field of environmental protection.[42]

The European Community, therefore, on the basis of the provisions just examined, enjoys an exclusive external competence in the commercial policy sector, with the limits highlighted by the ECJ in Opinion 1/94, while at the same time, it has a concurrent, albeit explicit, competence to conclude international agreements in the environmental field. Could these two distinct competences possibly collide and give rise to international controversies? The answer to such a question must be found by looking at the several Multilateral Environmental Agreements (MEAs) that the EC has concluded through the years and their potential conflict with the international free trade provisions enshrined in the GATT and in other international trade related treaties.

A great deal of attention has been devoted to this issue in the past few years in the legal literature and several MEAs have been identified whose provisions may possibly limit the freedom of international trade, such as for instance the 1987 Montreal Protocol on the protection of the ozone layer, the 1989 Basel Convention on the Transboundary Movement of Hazardous Waste and the 2000 Biosafety Protocol.[43] However, in practical terms no such controversy has ever been raised at international level and with time such agreements have been considered to be implicitly compatible with international trade provisions. In the same vein, some scholars have even been defining the trade and environmental treaties at hand as 'mutually supportive' ones.[44]

A contrast among treaty provisions pursuing trade and environmental goals, however, may also have relevance for the EC internal dimension. An example of this is represented by the issue of choosing the correct legal basis for the adoption of international treaties which pursue both commercial and environmental objectives simultaneously. Such an issue was at stake, for instance, in the ECJ

---

[42] See Dashwood, supra n. 3, p. 119ff.
[43] See in general on this topic Francioni, supra n. 40; Sands, P., *Principles of International Environmental Law*, 2nd edn (2003), p. 942ff.
[44] See Petersmann, in Pavoni, R., *Biodiversità e Biotecnologie nel diritto internazionale e comunitario* (2004), p. 219.

Opinion 2/00 on the EC's accession to the 2000 Cartagena Protocol on Biosafety, which supplemented the original 1992 Biodiversity Convention.[45]

After stating in general terms that 'under the system governing the powers of the Community, the choice of the legal basis for a measure, including one adopted in order to conclude an international agreement [...] must rest on objective factors which are amenable to judicial review', with regard to the specific issue of the correct legal basis in a case where a treaty pursues a twofold purpose, the European Court of Justice affirmed that the interpreter should try to identify the 'main or predominant purpose or component'. Based on this, it then held that the corresponding legal measure needed to give effect to such an agreement in the EC legal order had to be the appropriate legal basis foreseen by the EC Treaty for such a predominant aim, leaving aside the other incidental objective pursued by the said agreement.[46]

On the basis of such reasoning, the Court held that 'even if the control procedures set up by the Cartagena Protocol on Biosafety are applied most frequently, or at least in terms of market value preponderantly, to trade in living modified organisms, the fact remains that the Protocol is, in the light of its context, its aim and its content, an instrument intended essentially to improve biosafety and not to promote, facilitate or govern trade'. The Court therefore affirmed that the conclusion of the Cartagena Protocol on behalf of the European Community had to be founded on a single legal basis, specifically, albeit not exclusively, related to environmental policy, namely Article 175(1) EC Treaty. However, in practical terms, the ECJ also underlined that 'since the harmonization achieved at Community level in the Protocol's field of application covers in any event only a very small part of such a field, the Community and its Member States share competence to conclude the Protocol', which means that the EC and its Member States were to proceed with the adoption of the Cartagena Protocol through a 'mixed agreement'.[47]

As the case of the Cartagena Protocol shows, the conclusion of a 'mixed agreement' to which both the European Community and its Member States are Parties, is in fact a very useful solution in the framework of the EC legal order, and prevents possible controversies between the EC and its Member States in all cases of shared competence, such as in the environmental field.

In any case, in the 'trade and environment' sector, the most common situation which causes most controversies to arise is not related to what we have been discussing so far, but rather to the unilateral adoption of internal measures in the European legal context, either at the EC or Member State level, which pursues

---

[45] The 2000 Cartagena Protocol on Biosafety supplements the 1992 Biodiversity Convention and deals in particular with transboundary movement of living modified organisms resulting from modern biotechnology that may have adverse effects on the conservation and sustainable use of biological diversity.
[46] See ECJ Opinion 2/00, [2001] ECR I-9713, § 18.
[47] Ibid. § 22–3.

legitimate environmental protection objectives that may limit the free flow of international trade. This topic will be the focus of the subsequent analysis.

B. The WTO Context and the Relevant GATT Provisions and Case Law

The most common situation relating to the trade and environment issue and involving the European Community is represented by those controversies which originate at the international level from the unilateral adoption of internal measures in the European legal context, either at EC or Member State level, and which pursue legitimate environmental protection objectives that possibly conflict with pre-existing free trade obligations. In fact, every time that the EC or some of its Member States adopt a unilateral measure aimed at the protection of the environment, which may have an external dimension, the issue of the possible conflict with WTO and related treaties containing free trade obligations almost inevitably arises. In such a context, the WTO adjudicatory bodies will normally be called to address and possibly solve the question at hand, by trying to strike a balance between the opposed trade and commercial interests on the one side and the environmental or health protection goals on the other.

In such a context, therefore, we should try to determine whether, thus far, the WTO dispute settlement bodies have correctly interpreted and applied the relevant WTO provisions with respect to the conflicting environmental or health protection interests involved. To this end, let us start our analysis from the relevant GATT provisions and case law and then examine some relevant norms of two other trade treaties, namely the SPS and TBT Agreements, as well as the related case law, which will help us to better address the 'trade and environment' issue.[48]

The GATT Treaty has been, since 1947, one of the main driving forces behind the outstanding growth of the world economy in the second half of the twentieth century, characterized by the progressive liberalization of international trade achieved through a generalized abolition of non-tariff barriers and a trend towards the lowering and/or removal of tariff barriers to trade.

In 1994, at the conclusion of the Uruguay Round, the last of the several periodical rounds held by the GATT Contracting Parties in order to update the international rules governing global trade, the Parties decided to set up an international organization, namely the WTO, with the aim of administering the GATT Treaty and a whole series of other international trade agreements so as promote an increased undisturbed trade flow across the borders of the Contracting Parties worldwide.

---

[48] The following analysis of the GATT, TBT, and SPS relevant provisions and related case law with regard to the 'trade and environment' issue is drawn, with some amendments and integrations, from Montini, supra n. 40.

The main principle upon which the GATT is based is the principle of non-discrimination, enshrined in two clauses: (1) the 'most-favoured nation' clause (MFN clause), according to which all tariff concessions accorded to one State must be automatically extended to all other GATT Contracting Parties (Article I GATT); (2) the 'national treatment' clause (NT clause), on the basis of which imported goods, once they have entered into the national market of a given State, must not be subject to a less favourable treatment than national products (Article III GATT). Moreover, the MFN and NT clauses are supplemented by a general provision which prohibits all non-tariff restrictions on imported products (Article XI GATT).

On the basis of such provisions, the freedom of trade is the main goal and the main concern of the GATT Treaty. However, this does not mean that international trade must remain unrestricted in all circumstances. The GATT Treaty itself contemplates some circumstances in which trade in goods may be legitimately restricted in order to afford adequate protection to important interests of the Contracting Parties, provided that certain requirements are met. Such circumstances are listed in Article XX GATT, which is named 'General exceptions'.[49]

Pursuant to the provision of Article XX GATT, in order for a Party to justify the adoption of some national measures by giving priority to its national environmental interests over the general interest of the promotion of free and undisturbed international trade, two conditions must be fulfilled.[50] Firstly, it must be demonstrated that the national measures in question fall under one of the two following exceptions: (1) Article XX(b), which deals with national measures 'necessary to protect human, animal or plant life or health'; (2) Article XX(g), which refers to national measures 'relating to the conservation of exhaustible natural resources, if such measures are made effective in conjunction with restriction on domestic production or consumption'. Secondly, once ascertained that the national measures at stake fall under Article XX(b) or XX(g) exceptions, they must be assessed in light of the introductory clause of Article XX, the

---

[49] The rationale behind the 'general exceptions' listed in Article XX GATT has been aptly explained by E.U. Petersmann, who held that: 'the "general exceptions" in Article XX are designed to allow Contracting Parties to give priority to the "public policies" listed in Article XX over trade liberalisation by authorising trade restrictions necessary for the pursuit of overriding public policy goals, including protection of life, health and environmental resources'. See Petersmann, supra n. 40, at 29.

[50] The correct method of application of the 'general exceptions' contained in Article XX GATT for the justification of national measures aimed at the protection of the environment was summarized as follows by the Appellate Body (AB) in the *US Gasoline* case: 'In order that the justifying protection of Article XX may be extended to it, the measure at issue must not only come under one or another of the particular exceptions—paragraphs (a) to (j)—listed under Article XX; it must also satisfy the requirements imposed by the opening clauses of Article XX. The analysis is, in other words, two-tiered: first, provisional justification by reason of characterisation of the measure under XX—paragraphs (a) to (j); second, further appraisal of the same measure under the introductory clauses of Article XX.'

so-called *chapeau*, which states that they can be held compatible with the GATT provided that: 'such measures are not applied in a manner which would constitute a means of arbitrary or unjustifiable discrimination between countries where the same conditions prevail, or a disguised restriction on international trade'.[51]

C. The Application of Article XX(b) GATT

An analysis of the relevant GATT case law on trade and environment ought to start from the exception contained in Article XX(b) GATT. In this respect, the first case decided after the institution of the WTO which involved the application of the exception contained in Article XX(b) GATT was the *US Gasoline* case (1996).[52] The *US Gasoline* case arose from a US regulation which imposed on foreign gasoline refiners wishing to export their products into the US territory stricter environmental standards than those applicable for US national refiners. The Panel, when examining the national measure under Article XX(b), focused in particular on the 'necessity' of the US measure. In this way, in fact, the Panel recalled the interpretation of the term 'necessary' already given by some previous GATT Panels in the *US Section 337 case* (1989) and in the *Thai Cigarettes case* (1990) and stated that, with regard to the US measure under scrutiny: 'If there were consistent or less inconsistent measures reasonably available to the United States, the requirement to demonstrate necessity would not have been met'.[53]

The Panel therefore found the US measure unjustifiable under Article XX(b), since there were alternative measures less restrictive for international trade which could have been adopted by the US authorities. The case was then appealed before the WTO Appellate Body (AB).[54] The AB did not explicitly deal with the Article XX(b) exception and with the interpretation of the term 'necessary' contained therein, but rather decided the case purely on the basis of Article XX(g). However, from an *obiter dictum* of the decision, it seems that the AB largely wished to endorse the line of reasoning proposed by the Panel, according to which a national measure could be considered 'necessary' within the meaning of Article XX(b) only if it was possible to demonstrate that there were no alternative measures reasonably available to the State that could achieve the aim sought with a lower impact on international trade.[55]

In the subsequent environmental case decided under Article XX GATT, namely the *Shrimps/Turtles* case (1998),[56] which dealt with a US regulation

---

[51] See Charnovitz, S., 'Exploring the Environmental Exceptions in GATT Article XX', (1991) *Journal of Environmental Law* 5.
[52] See *US Gasoline* case, Report of the Panel, WT/DS2/R, in (1996) 35 *ILM* 274.
[53] See *US Gasoline* case, Report of the Panel, supra n. 52, § 6.24.
[54] See *US Gasoline* case, Report of the Appellate Body, WT/DS2/AB/R, in (1996) 35 *ILM* 605.
[55] See *US Gasoline* case, Report of the Appellate Body, supra n. 50, 620.
[56] See *Shrimps/Turtles* case, Report of the Panel, WT/DS58/R and Report of the Appellate Body, WT/DS58/AB/R.

imposing certain fishing technologies to prevent accidental take of sea turtles to fishing activities occurring outside the jurisdiction of the United States, the US authorities sought justification both under Article XX(b) and XX(g), but the Panel and the Appellate Body only addressed the issues under the Article XX(g) exception, and did not provide any further interpretative contribution to the interpretation and application of the Article XX(b) exception. Therefore, the case will be dealt with below with reference to the Article XX(g) exception.

The application of the Article XX(b) exception was addressed again in the *EC Asbestos* case (2001).[57] The case arose from a French national regulation which banned 'the manufacture, import, domestic marketing, exportation, possession for sale, offer, sale and transfer under any title whatsoever of all varieties of asbestos fibres or any product containing asbestos fibres' for the protection of workers and consumers. The French regulation did, however, allow a limited exception to the general ban in specific cases when no substitute for materials, products, or devices containing chrysotile fibres was available.

Canada, which prior to the adoption of the national measure at stake was a major exporter of asbestos fibres and products to France, challenged the French Decree before the WTO Dispute Settlement Body (DSB), claiming that the Decree: (1) constituted a technical regulation covered by the Agreement on Technical Barriers to Trade (TBT Agreement) and was incompatible with various provisions of the TBT Agreement; (2) amounted to a ban incompatible with Article III(4) GATT and was not justifiable under the exception contained in Article XX(b) GATT.

As regards the claim concerning the compatibility of the French measure with the GATT provisions, the issues under scrutiny were the following: firstly, it had to be determined whether 'chrysotile asbestos fibres and products' and 'PCG fibres and products' were 'like products' within the meaning of GATT Article III(4)[58] and secondly, it had to be assessed whether the French measure could be justified under the Article XX(b) exception.

For the purpose of the present analysis we will deal here solely with the latter issue. In this respect, the Panel firstly ascertained that, on the basis of the scientific evidence available, the asbestos fibres and products constituted a risk for public health and therefore the French regulation fell within the range of policies designed to protect human life or health, within the meaning of GATT Article XX(b), and secondly, it concretely assessed the justifiability of the national measure by applying the 'necessity principle', as already developed by other

---

[57] See *EC Asbestos* case, Report of the Panel, WT/DS/135/R, and Report of the Appellate Body, WT/DS/135/AB/R.

[58] See Article III(4) of the GATT which reads, in the relevant part: 'The products of the territory of any Member imported into the territory of any other Member shall be accorded treatment no less favourable than that accorded to like products of national origin in respect of all laws, regulations and requirements affecting their internal sale, offering for sale, purchase, transportation, distribution or use'.

Panels and the AB in previous cases.[59] In doing so, the Panel analysed in particular whether other measures 'consistent or less inconsistent' with the prescriptions of the GATT were 'reasonably available' to France and found that, in light of the 'high degree' health policy objectives sought by France, no measures existed which could be considered a 'reasonably available' alternative to the general ban adopted.[60] Therefore, the Panel concluded that the French measure was justifiable under the exception of Article XX(b) GATT, after having verified that it also satisfied the requirements of the *chapeau*.[61]

The Panel's decision was then reviewed by the AB. As to the issue of the justification of the French measure under Article XX(b) GATT, in particular, the AB substantially upheld the Panel's finding that the French Decree could be justified under Article XX(b), by underlining that since 'the chosen level of health protection by France was a halt to the spread of asbestos-related health risks', the general ban on asbestos products was clearly designed to achieve the objective sought and there was no alternative measure 'reasonably available' to France 'that would achieve the same end and that is less restrictive of trade than a prohibition'.[62]

The cases just analysed above show that the Panels and the AB, when judging the possibility of justifying a national measure under an Article XX(b) exception have always (more or less consistently) applied the same sort of balancing instrument, which can be named the 'necessity principle', as I have described in greater detail elsewhere.[63]

In brief, on the basis of such a balancing instrument, in the AB's view, 'in order to determine whether a measure is necessary it is important to assess whether consistent or less inconsistent measures are reasonably available'. The rationale behind such an instrument is therefore the following: the national measure must try to achieve the legitimate objective sought by posing the least possible degree of restriction on international trade. If, however, there are no alternative measures 'reasonably available' to achieve the same degree of environmental or public health protection, as for instance in the *EC Asbestos* case, even quite a burdensome restriction to trade, such as a general ban on certain goods, may be held justifiable under Article XX(b).

A similar line of reasoning was followed by the Panel and the AB in the recent *Retreaded Tyres* case (2007). The case originated from a complaint raised by the European Community concerning the consistency with the GATT of certain measures imposed by Brazil on the importation and marketing of retreaded tyres,

---

[59] See for instance *US Gasoline* case and *Shrimp/Turtles* case, supra nn. 48, 50, and 52.
[60] See *EC Asbestos* case, Report of the Panel, supra n. 57, § 8.208–8.223.
[61] Ibid. § 8.241.
[62] See *EC Asbestos* case, Report of the Appellate Body, supra n. 57, § 164–175.
[63] See Montini, M., 'The Necessity Principle as an Instrument to Balance Trade and the Protection of the Environment', in Francioni, F. (ed.), *Environment, Human Rights and International Trade* (Hart Publishing, 2001), p. 135ff.

most notably consisting in an import ban and various other related measures. The Panel, with regard to the necessity of the national measure at stake under the Article XX(b) exception, held that:

the necessity of a measure should be determined through 'a process of weighing and balancing a series of factors', which usually includes an assessment of the following three: the relative importance of the interests or values furthered by the challenged measure, the contribution of the measure to the realization of the ends pursued by it and the restrictive impact of the measure on international commerce.

Moreover, it also added that 'Once all those factors have been analysed, a comparison should be undertaken between the challenged measure and possible alternatives'.

On the basis of this line of reasoning, the Panel found that the import ban on retreaded tyres could be considered as being 'necessary to protect human, animal or plant life or health' and therefore could be provisionally justified under Article XX(b). However, the Panel then found that the import ban did not satisfy the requirements of the *chapeau* of Article XX, insofar as it was concretely applied in a manner that constituted both a means of unjustifiable discrimination and a disguised restriction on international trade.

The AB, on appeal, affirmed that the rationale for the interpretation of the term 'necessary' involves 'a process of weighing and balancing a series of factors which prominently include the contribution made by the compliance measure to the enforcement of the law or regulation at issue, the importance of the common interests or values protected by that law or regulation, and the accompanying impact of the law or regulation on imports or exports'. In practical terms, the AB held that in order to determine whether a measure is 'necessary' within the meaning of Article XX(b) GATT, all the relevant factors must be examined, but in particular the focus must be on the extent of the contribution that such a measure may give to the achievement of the desired objective and on its trade restrictiveness, in light of the importance of the interests or values at stake. Moreover, if such an analysis brings us to the preliminary conclusion that the measure seems to be 'necessary', this provisional result must be confirmed by comparing the measure with its possible alternatives, which may be less trade restrictive while providing an equivalent contribution to the achievement of the objective pursued.

As for the analysis to be conducted under the *chapeau* of Article XX, the AB stated that the determination of whether the application of a national measure amounts to an arbitrary or unjustifiable discrimination involves an analysis that relates primarily to the cause or the rationale of the discrimination. On this basis, the AB, recalling some of its previous case law on the matter, restated that the analysis of whether the application of a measure results in arbitrary or unjustifiable discrimination should focus on the cause of the discrimination or on the rationale put forward to explain its existence, rather than on the effects of the discrimination. On the basis of such a premise, the Appellate Body, with respect to the analysis of the 'necessity' of the import ban imposed by Brazil on retreaded tyres under Article XX(b) of the GATT, upheld the Panel's finding that the import ban could be considered 'necessary' within the meaning of Article XX(b) and was thus provisionally justified under that provision. However, despite reversing some of the Panel's findings regarding the correct interpretation and application of the *chapeau* of Article XX, the AB finally confirmed that the import ban could not be justified under Article XX of the GATT.

D. The Application of Article XX(g) GATT

In the already mentioned *US Gasoline* case (1996),[64] the US national measure at stake was also examined under Article XX(g) after being analysed under Article XX(b). In this respect, the Panel, when applying the Article XX(g) exception, did not depart from the interpretation of the terms 'relating to' the protection of an exhaustible natural resource as 'primarily aimed at' such protection, as well as from the interpretation of the term 'made effective in conjunction with restrictions on domestic production or consumption' as 'primarily aimed at rendering effective these restrictions', which had been constantly proposed by the previous GATT Panels.[65]

The Panel's decision, however, was appealed before the AB, which on one hand confirmed that the interpretation of the term 'relating to' as meaning 'primarily aimed at' the conservation of an exhaustible natural resource was correct, but, that on the other, when interpreting the term 'made effective in conjunction with restrictions on domestic production or consumption' noted that it seems to require not so much that the national measures must be 'primarily aimed at rendering effective these restrictions', as had been held by the Panel, but rather that 'the measures concerned impose restrictions, not just in respect of imported gasoline but also with respect to domestic gasoline'.[66] Therefore, the AB concluded that the US national measure was provisionally justifiable under the Article XX(g) exception. However, the AB finally found that the measure, as it was concretely applied, did not satisfy the requirements of the *chapeau*, insofar as it constituted an 'unjustified discrimination' and a 'disguised restriction to international trade'. As a consequence, the said measure could not be justified under Article XX(g) GATT.[67]

In the *Shrimps/Turtles* case (1998),[68] as mentioned above, the US regulation under scrutiny was examined only under the Article XX(g) exception, although both Article XX(b) and XX(g) had been invoked by the defendant. In this case,

---

[64] See *US Gasoline* case, supra nn. 52 and 54.
[65] See *US Gasoline* case, Report of the Panel, supra n. 52, § 6.35–6.42.
[66] See *US Gasoline* case, Report of the Appellate Body, supra n. 54, at 22.
[67] Ibid. at 29.
[68] See *Shrimps/Turtles* case, supra n. 56.

the issue of the interpretation and application of the Article XX(g) exception was not explicitly addressed by the Panel. The AB, however, when examining the term 'relating to' the conservation of an exhaustible natural resource, although formally not departing from the traditional interpretation of the term as referring to measures 'primarily aimed at' the conservation of an exhaustible natural resource, in practice gave some interesting suggestions on how to better address the 'relationship between the general structure and design of the measure at stake and the policy goal it purports to serve'.[69] The AB, in fact, found that the national measure at stake was justifiable under the Article XX(g) exception insofar as it was 'not disproportionately wide in its scope and reach in relation to the policy objective of the protection and conservation of sea turtles species' and 'the means [we]re in principle reasonably related to the ends'.[70]

With the above-mentioned statement, the Appellate Body seems, in fact, to propose the adoption of a new interpretative test according to which the possibility of justifying a national measure under Article XX(g) ought to depend not so much on the assessment that the measure at stake is 'primarily aimed' at the conservation of an exhaustible natural resource, but rather on the evaluation that the measure is not 'disproportionately wide in its scope and reach in relation to the policy objective' or in other words is 'reasonably related to the ends'. The test proposed by the AB thus seems to introduce a sort of necessity and proportionality dimension into the evaluation of a national measure under the Article XX(g) exception, which recalls the application of the balancing instrument of the 'necessity principle' already seen above with reference to Article XX(b).

We have now analysed how the 'trade and environment' issue has been addressed and solved in the most relevant case law under the GATT. In the next two paragraphs, the analysis will shift to two other important WTO multilateral agreements related to trade in goods, namely the SPS and the TBT, in order to verify how the WTO dispute settlement authorities have approached the issue of the balance between conflicting trade end environmental interests in the framework of such treaties.

E. The Case Law on the SPS Agreement

The SPS Agreement (*Agreement on the Application of Sanitary and Phytosanitary Measures*) contains a series of rules and procedures for the definition and application of national sanitary and phytosanitary standards by its Members, which are all WTO Parties. In general terms, the SPS Agreement sets a preference for the use of internationally agreed standards, whenever possible. However, it also recognizes the possibility for Members to adopt, in certain circumstances, national sanitary and phytosanitary measures based on more restrictive standards

---

[69] See *Shrimps/Turtles* case, Report of the Appellate Body, supra n. 56.
[70] Ibid. § 141.

than those agreed at international level, provided that such standards are either based on 'scientific justification' or determined on the basis of a 'risk assessment procedure' conducted pursuant to the provisions of Article 5 of the SPS Agreement. In this respect, it is noteworthy that such a 'risk assessment procedure' incorporates the proportionality principle (Article 5.6) and the precautionary principle (Article 5.7).

The first circumstance in which the application of the SPS Agreement was invoked before the WTO dispute settlement authorities was the *EC Hormones* case (1998).[71] The case arose from a ban imposed by the European Community on the production, sale, and importation of meat and meat products taken from cattle treated with hormones. The ban on import, in particular, blocked US and Canadian import of meat into Europe and, after the failure to solve the dispute through negotiations with the Parties, the case ended up before the WTO dispute settlement authorities.

For the purpose of the present analysis, it is necessary to deal extensively with the outcome of the *EC Hormones* case, which essentially focused on the issue of the correct application of the precautionary principle, as enshrined under Article 5.7 of the SPS Agreement, and ended up with a condemnation of the ban on hormones, meat, and meat products treated with hormones adopted by the European Community. It is important to stress, in this respect, that according to the procedure established under the SPS Agreement as interpreted by the WTO Appellate Body, when a Member wishes to adopt a national sanitary or phytosanitary measure based on more restrictive standards than those agreed at international level, the following requirements must be fulfilled: (1) firstly, all national measures aimed at the protection of human, animal, or plant health or life must be based on a 'risk assessment' based on available scientific evidence; (2) secondly, when determining the appropriate level of SPS protection, Members should 'take into account the objective of minimising negative trade effects'; (3) thirdly, the national measures chosen should not be 'more trade-restrictive than required to achieve their appropriate level of SPS protection, taking into account technical and economic feasibility'; (4) fourthly, in case full scientific certainty does not exist, Members may provisionally adopt 'precautionary measures' which must be temporary and should be reviewed as soon as possible.

On the basis of the four requirements described above, it can be argued that in order to determine whether a national measure setting higher standards than those internationally agreed upon can be upheld under the SPS Agreement, it must be previously assessed that: (1) the national measure is necessary to afford adequate protection to human, animal, or plant health or life and (2) the national measure is not more trade-restrictive than required to achieve the aim sought. This is, in fact, nothing but another application of the 'necessity principle'

---

[71] See *EC Hormones* case, Reports of the Panel, WT/DS26/R/USA and WT/DS26/R/CAN; Report of the Appellate Body, WT/DS26-DS48/AB/R.

composed by the necessity and proportionality tests which we have already seen in the case law under the GATT.

More recently, the application of the SPS Agreement has been invoked again before a WTO Panel in the *EC Biotech* case (2006).[72] In this instance, the three major exporters of agricultural and food products containing genetically modified organisms (GMOs), namely the United States, Canada, and Argentina, had filed three separate complaints against an alleged moratorium by the EC on the approval of GMOs during the period from October 1998 to August 2003, as well as against some of the EC Member States' national bans on GMOs and GM foods.[73]

On the first question, in particular, the Panel found that the EC had in fact applied a de facto moratorium on the approval of GMOs. However, it held that the moratorium did not consist in an SPS measure within the meaning of the SPS Agreement, although it certainly affected the operation of the EC provisions on the approval of GM products, rather it resulted in a less relevant '*failure to complete individual approval procedures without undue delay*'. This 'procedural solution' to the question at stake in fact enabled the Panel to refrain from addressing several very relevant legal issues, including the one relating to the consistency with WTO trading rules (or more precisely with the SPS provisions) of the approval procedures established by the relevant EC provisions, namely those of Directives 90/220 and 2001/18 and Regulation 258/97, which provided for a product-by-product assessment requiring scientific consideration of various potential risks.[74]

As regards the second question, the Panel firstly held that the contested safeguard measures taken by some EC Member States (namely Austria, Belgium, France, Germany, Italy, and Luxembourg) fell within the scope of the SPS Agreement and, after a careful analysis, concluded that such safeguard measures violated Articles 5.1 and 2.2 of the SPS Agreement. The Panel based its decision particularly on that fact that since the EC Scientific Committee had comprehensively evaluated the potential risks to human health and/or the environment prior to the granting of Community-wide approval of the relevant GM products and had provided a positive opinion in this regard, Member States could not maintain that in the case at stake the relevant scientific evidence was insufficient to perform a risk assessment, in which case they might have been allowed to have recourse to provisional measures under Article 5.7 of the SPS Agreement.[75]

The Panel decision in the *EC Biotech* case was therefore only partially unsatisfactory for the European Community, which was not actually condemned on the merits of its GMOs approval procedures and therefore decided not to appeal

---

[72] See *EC Biotech* case, Reports of Panel, WT/DS291/R, WT/DS292/R, WT/DS293/R.
[73] For a basic comment on the *EC Biotech* case see Lenzerini, F. and Montini, M., 'The Activity of the World Trade Organization (2006)', (2006) 16 *The Italian Yearbook of International Law* (2007).
[74] See *EC Biotech* case, supra n. 72, § 8.6–8.7.
[75] See *EC Biotech* case, supra n. 72, § 8.9–8.10.

the Panel's decision before the Appellate Body. However, from a legal analysis point of view, the outcome of the case was very disappointing, insofar as the Panel missed the opportunity to address and clarify the most critical environmental and health implications related to the application of national measures taken by some WTO Parties to respond to their concerns raised by the import of GMOs.

F. The Case Law on the TBT Agreement

The TBT Agreement (*Agreement on Technical Barriers to Trade*) deals with the application of technical measures by Members, which may negatively interfere with international trade. In general terms, the TBT Agreement encourages Members to adopt internationally agreed technical standards whenever possible. However, Members are not totally prevented from taking national measures necessary to protect human, animal, and plant life or health or the environment. Quite on the contrary; in principle, each Member has the right to determine the level of protection that it deems more appropriate, provided that 'technical regulations are not prepared, adopted or applied with a view to or with the effect of creating unnecessary obstacles to international trade'. Moreover, 'technical regulations shall not be more trade restrictive than necessary to fulfil a legitimate objective, taking into account the risk non-fulfilment would create'.[76]

As one can see, the TBT Agreement fully embodies the necessity principle by imposing upon Members wishing to adopt their own national technical standards the burden to ensure that their measures satisfy both the necessity test, in the sense that they fulfil a legitimate objective, such as the protection of the environment, and the proportionality test, in the sense they are not 'more trade restrictive than necessary to fulfil a legitimate objective' nor adopted or applied 'with a view to or with the effect of creating unnecessary obstacles to trade'.

Up to now, the TBT Agreement has been invoked twice in WTO controversies related to 'trade and environment'. In the first case, namely the *US Gasoline case* (1996),[77] the issue of the compatibility of the national measure at stake with the TBT Agreement was not analysed by the Panel and the AB. Conversely, in the second case, namely the *EC Asbestos* case (2001), the Panel and the AB partially addressed the Canadian claim regarding the alleged violation of some provisions of the TBT Agreement by the French ban, but unfortunately did not deal with the most interesting issue of the justifiability of the national provision under Article 2(2) of the TBT Agreement. In fact, the Panel and the AB limited themselves to investigating whether the national measure under scrutiny was a 'technical regulation' within the meaning of Annex 1(1) to the TBT Agreement.[78]

---

[76] See Article 2.2 TBT Agreement, reprinted (1994) 33 ILM 28ff.
[77] See *US Gasoline* case, supra nn. 52 and 54.
[78] See *EC Asbestos* case, supra n. 57.

Case Study on EC External Relations on Trade and the Environment   157

Since the national measure at stake in the *EC Asbestos* case was not examined under Article 2(2) of the TBT Agreement, a basic question remained unanswered: in the absence of any case law on the topic, is it possible to envisage a further extension of the same *acquis* elaborated by the Panels and the AB with regard to Article XX(b) and XX (g) GATT and Article 5 SPS Agreement, to the realm of Article 2(2) TBT Agreement?

Such an extension was in fact envisaged by Canada in its pleadings in the *EC Asbestos* case. However, neither the Panel nor the AB specifically addressed the issue. This notwithstanding, the claimed extension seems to be perfectly possible, since the analysis conducted above has shown that the 'purpose' of the four treaty provisions considered (namely Article XX(b) and XX(g) GATT, Article 5 SPS Agreement, Article 2(2) TBT Agreement) is essentially the same. Of course, this does not mean that the four provisions under scrutiny are identical, as each one clearly differs from the others with regard to its specific scope. However, the existence of a different specific scope, if counterbalanced by the presence of the same 'purpose', should not prevent the four provisions from being considered together as an expression of the 'necessity principle'. Therefore, in more concrete terms, nothing should preclude the four provisions from having substantially the same role as a 'balancing instrument' to be applied by the WTO dispute settlement authorities in order to find, on a case-by-case basis, the best solutions in 'trade and environment' disputes.[79]

G. Remarks on the EC's Behaviour in the Context of the WTO Case Law

Now that we have analysed the types of reasoning and instruments the WTO dispute settlement bodies have devised in order to balance competing trade and environmental interests, let us try in this final paragraph of the present section to propose some brief remarks on the essence of the EC's position in the WTO context, by throwing a quick glance at the EC's behaviour in the major relevant trade and environmental cases adjudicated in the framework of the WTO dispute settlement regime.

In this respect, if one looks at the already mentioned cases which involved the European Community as an actor and which are relevant for the 'trade and environment' debate, such as the *EC Hormones* case, the *EC Asbestos* case, and the *EC Biotech* case, it is possible to argue that in this sector a sort of 'transatlantic clash' has emerged between the positions of the two main players, namely the EC and the US. In fact, in all three cases cited above, the EC tried to impose on all its trade partners an 'environmentally friendly' solution to an emerging

---

[79] In this respect, see Montini, M., *La necessità ambientale nel diritto internazionale e comunitario* (2001). See also Montini, M., 'The Necessity Principle as an Instrument to Balance Trade and the Protection of the Environment', supra n. 63.

environmental or health problem, which affected trade in goods and thus limited international commerce. On the other hand, in all those circumstances, the US challenged this 'protective' approach, arguing that the risks for the environment or public health were not so high as maintained by the EC, and that, in any case, the unilateral measures taken by the EC in order to protect the interests at stake were excessive with respect to the objective sought and thus posed an unnecessarily high burden on trade.

In all of these cases, irrespective of their final outcome, as decided by the WTO dispute settlement bodies, the EC position reflected the very strong perception that once an 'essential interest' such as the protection of the environment is under a serious challenge, the norms on the freedom of international trade must give way to such an interest and retreat. This approach is certainly not shared by the US, nor by several other WTO Parties, and probably reflects the different way of approaching environmental problems on the two sides of the Atlantic Ocean. At this point, the interpreter is called to try and determine the origin of such a distinctive EC approach. In order to do so, it seems that two essential questions need to be answered. The first relates to whether the EC approach has been influenced by the provisions of the EC Treaty. The second, which logically follows the first, refers to whether such an approach resembles the EC 'internal' case law on 'trade and environment'.

The answer to the first question must take into account the fact that in the framework of the EC Treaty, the protection of the environment is given a paramount role. The objective of integrating environmental requirements in the definition and implementation of all other EC policies and activities is in fact stated in very clear terms in one of the general provisions of the EC Treaty, namely Article 6. Moreover, as far as the EC's 'external' behaviour in the environmental field is concerned, Article 174(1) needs to be considered, according to which EC environmental policy must contribute inter alia to the promotion of 'measures at international level to deal with regional or worldwide environmental problems'.

The provisions cited above, though they might partially explain the proactive role played by the EC in the international environmental arena, do not directly influence the adjudication of EC 'internal' trade and environmental disputes (which relate to the legitimacy under the applicable EC Treaty provisions of the unilateral national measures which might be taken by an EC Member State in order to afford a high level of protection to one of its environmental interests and which conflict with its trade obligations posed by the same EC Treaty). In this respect, attention must be drawn instead to Article 30 EC Treaty, which refers to the 'protection of health and life of humans, animals and plants' as a possible cause of exception to the general rule on the free movement of goods contained in Article 28 EC Treaty. Moreover, the jurisprudence of the European Court of Justice has supplemented this cause of exception with a more sophisticated one, known as the 'mandatory requirements' doctrine, first stated in the celebrated

Cassis de Dijon (1979) case and again in several other cases, among which special mention must be made of the *Used Oils* case (1984), when a direct reference to the protection of the environment was first inserted in the context of the 'mandatory requirements' doctrine.[80]

In practice, the two types of exceptions provided by Article 30 EC Treaty and by the 'mandatory requirements' doctrine elaborated by the ECJ, notwithstanding their basic differences, ultimately pursue the same objective of rendering more flexible the application of EC Treaty rules to the free movement of goods, thus enabling EC Member States to maintain an adequate protection of their fundamental national interests in the environmental field, without dismissing their free trade obligations under the EC Treaty.

On the basis of the above remarks, and in answer to the first question posed above, it can be said that the EC's 'protective' approach observed in the WTO context is clearly influenced by EC Treaty provisions, insofar as in this context, which in many respects recalls the GATT provisions, the protection of the environment is substantially accorded a greater role with respect to the free movement of goods, compared to that assigned to it in the GATT context, although in general terms environmental protection still represents an exception to the provisions on free trade in goods contained in the EC Treaty.

As to the second, and probably most interesting, question posed above, the most relevant EC case law in the 'trade and environment' context, including for instance some of the older cases such as the *Danish Bottles* case (1986), the *Walloon Waste* case (1992), as well as some of the most recent cases such as the *Danish Bees* case (1998) and the *Preussen Elektra* case (2001), which obviously cannot be properly analysed here,[81] show that the ECJ, when adjudicating the legitimacy of a unilateral national measure taken by a Member State vis-a-vis the norms on the free movement of goods, has constantly made use of a balancing instrument composed of the necessity and proportionality tests, which closely resembles the one adopted by the WTO dispute settlement authorities in similar cases already mentioned above.[82]

In fact, if one compares the main features of the necessity and proportionality tests as they emerge both in the WTO and EC case law on 'trade and environment', it seems that a very similar reasoning is used in both contexts. As far as the WTO case law is concerned, for instance, under the necessity test the dispute settlement bodies normally ascertain whether a national measure taken by a WTO Party is apt and suitable to adequately protect a certain important interest

---

[80] See Case 120/78 *Cassis de Dijon* [1979] ECR 649; Case 240/83 *Used Oils* [1984] ECR 531.
[81] See Case 302/86 *Danish Bottles* [1988] ECR 4607; Case C-2/90 *Walloon Waste* [1992] ECR I-4431; Case C-67/97 *Danish Bees* [1998] ECR I-8033; Case C-379/98 *Preussen Elektra* [2001] ECR I-2099. For a detailed analysis of such cases see Montini, M., 'Trade and Environment: An Analysis of the Conflicting Interests at Stake in the Case-Law of the European Court of Justice', (2/2003) *Czech Journal of Environmental Law (Ceske Pravo Zivotniho Prostredi)* 75ff.
[82] See Montini, M., 'The Necessity Principle as an Instrument to Balance Trade and the Protection of the Environment', supra n. 63, p. 135ff.

of such a Party, whereas under the proportionality test they try to determine whether the national measure chosen by the said Party exceeds what appears to be indispensable to reach the desired objective.

However, in the most recent cases decided by the WTO dispute settlement bodies, as shown above, a new trend is emerging which tends to depart from a rather formalistic approach and moves towards a more flexible approach. In broader terms, the older approach focused primarily on the indispensability and unavoidability of the unilateral measure at stake, by requiring the WTO Party wishing to adopt a certain measure to demonstrate that the chosen measure was the only available means to achieve the aim sought and that no alternative measures existed which could achieve the same result with a lower impact on international trade. Quite differently, the new approach seems to shift the focus towards the reasonableness of the unilateral measure at hand. In fact, thanks particularly to the AB jurisprudence, some of the most recent WTO decisions—while not abandoning the traditional line of reasoning according to which a unilateral measure taken by a Party must try to achieve the objective sought by posing the lowest possible burden on international trade—apparently consider that a more flexible approach focusing on the 'reasonableness' of the measure under scrutiny, is now deserved. This implies that in certain circumstances, when no alternative measures are 'reasonably available' to achieve the same degree of environmental or health protection, even quite a burdensome restriction posed on international commerce such as a general ban on the import of some kind of goods, may be held justifiable under the GATT/WTO provisions.

In quite similar terms, an analysis of the most relevant EC case law on trade and environment shows that the necessity test is normally used by the ECJ in order to determine whether a national measure taken by a Member State is suitable and apt in general and abstract terms to reach the desired objective, whereas under the proportionality test, the Court normally determines the reasonableness of the measure at stake with reference to the objective sought. Moreover, in the framework of the proportionality test, the existence of an alternative measure which was concretely and reasonably available to the State and which could have achieved the same result by posing a lower burden on intra-community trade, may be taken into account by the Court. However, such an element is not the object of a self-standing evaluation, but rather represents an additional element used by the ECJ to take a decision on the reasonableness of the national measure under scrutiny.[83]

On the basis of the brief remarks just made, therefore, it is possible to conclude that the EC internal case law in this field seems to have greatly influenced the EC's behaviour in the 'trade and environment' issues adjudicated in the WTO context and might also have partially influenced the reasoning of the WTO dispute settlement bodies, as demonstrated by the various analogies which can be

---

[83] See Montini, supra n. 81, p. 99.

4. Case Study on EC External Relations on Climate Change

A. Introduction on the International Climate Change Regime

The climate change issue arose in the late eighties, once scientific evidence started to suggest that the progressive increase in the concentration of greenhouse gases in the atmosphere may be contributing to a large extent to the 'greenhouse effect', which is in turn causing global warming and giving rise more generally to the so-called climate change phenomenon. To promote comprehensive and reliable studies on the climate change issue, in 1988 the UNEP and the WMO established the Intergovernmental Panel on Climate Change (IPCC). The IPCC was requested to provide scientific evidence upon which the international institutional and legal efforts to efficiently tackle climate change could be based.[84]

The first IPCC Report, published in 1990, confirmed that a rise in the temperature was already occurring and warned about the possible risks associated with this phenomenon. The IPCC findings, in fact, paved the way for a rapid negotiation of the first international legal instrument related to climate change, the UN Framework Convention on Climate Change (UNFCCC), which was signed at the 1992 Rio Conference on Environment and Development (UNCED).

The Framework Convention pursues a general commitment of greenhouse gas stabilization in the atmosphere over the long term and contains the basic principles and rules upon which the climate change legal regime is based, including specific provisions on measurement and reporting on greenhouse gas emissions. The UNFCCC was meant to represent the starting point for future stringent actions to be contained in subsequent legal instruments. Shortly after the entry into force of the Framework Convention, negotiations began for drafting a protocol with more precise commitments for the most industrialized countries around the world, which had already been listed in Annex I to the Convention.

---

[84] On the climate change regime in general see Yamin, F. and Depledge, J., *The International Climate Change Regime*, (Cambridge University Press, 2004); Victor, D., *The Collapse of the Kyoto Protocol and Struggle to Slow Global Warming*, (Princeton University Press, 2004); Verheyen, R., *Climate Change Damage and International Law*, (Martinus Nijhoff, 2005); Freestone, D. and Streck, C., *Legal Aspects of Implementing the Kyoto Protocol Mechanisms*, (Oxford University Press, 2005); Bothe, M. and Rehbinder, E., *Climate Change Policy* (Eleven International Publishing, 2005); Metz, B. and Hulme, M., (eds.), *Climate Policy Options post 2012: European Strategy, Technology and Adaptation after Kyoto* (Earthscan, 2005); Peeters, M. and Deketelaere, K. (eds.), *EU Climate Change Policy: The Challenge of New Regulatory Initiatives* (Edward Elgar, 2006); Douma, W. Th., Massai, L., and Montini, M. (eds.), *The Kyoto Protocol and Beyond: Legal and Policy Challenges of Climate Change*, (T.M.C. Asser Press, 2007).

The result of these is contained in the Kyoto Protocol on Climate Change, concluded in 1997 at the Third Conference of the Parties to the UNFCCC. The Kyoto Protocol, besides setting the duty upon Annex I Parties to draft appropriate policies and measures in the economic sectors that may have a relevant influence on the increase of GHG emissions, also imposes specific quantified emission limitation and reduction commitments to be achieved within a specific timeframe, that is during the first commitment period (2008–2012), with respect to their 1990 GHG emissions levels.

The emissions targets for the most industrialized countries, which are listed in Annex B to the Kyoto Protocol, may be achieved through a combination of national policies and measures coupled with the use of three economic instruments, the so-called 'flexibility mechanisms', namely Joint Implementation, the Clean Development Mechanism, and Emissions Trading.

The flexibility mechanisms are designed to reduce the marginal costs of GHG emission reductions by implementing projects that reduce emissions or remove carbon from the air in other Annex I Parties (Joint Implementation) or in non-Annex I Parties (Clean Development Mechanism), or alternatively by acquiring emission reduction units from other Annex B Parties (Parties with quantified emission limitation and reduction commitments) on the market (Emission Trading).

The Kyoto Protocol also establishes detailed provisions on the development of national inventories of GHG emissions as well as on reporting to the UNFCCC Secretariat. The information submitted by the Annex I Parties is then reviewed by ad hoc 'Expert Review Teams', on the basis of the guidelines adopted by the Meeting of the Parties to the Protocol. Finally, the information contained in the annual inventories may trigger the compliance mechanism which has been meanwhile established to tackle non-compliance with the commitments stemming from the Kyoto Protocol in a 'non adversarial' but effective way.[85]

B. The EC and the International Climate Change Regime

Since its inception, the EC has always been an active participant in the international climate change regime and, in fact, has constantly promoted not only the drafting and signature of the UNFCCC and the Kyoto Protocol, but also their ratification and effective implementation by the Parties, so as to radically tackle the climate change emergency.

The EC is committed to promoting actions and measures with an external relevance to deal with regional or worldwide environmental problems on the basis of Article 174(4) EC Treaty, which identifies the international dimension of EC

---

[85] On the Kyoto Protocol compliance regime see Montini, 'M., The Compliance Regime of the Kyoto Protocol', in Douma, W. Th., Massai, L., Montini, M., *The Kyoto Protocol and Beyond*, p. 95ff.

activities in the environmental field as one of the fundamental objectives of EC environmental law.

Within such a framework, the climate change activities and relevant legal acts have become in recent years one of the most dramatic priorities for EC external relations policy in the field of environmental law, as is demonstrated by the amendment made to the relevant provision of Article 174(4) EC Treaty, made in the recent 2007 Lisbon Treaty, which updates the present text of the EC Treaty. The said amendment, in fact, adds a specific reference to the climate change sector as the major reference subject matter with external relevance for the European Community.

From the internal EC point of view, the existence of such a favourable context for the activities of the EC institutions in the field of climate change—in order to work effectively—must be coupled with corresponding provisions within the specific legal instruments which enable a full and meaningful participation of the European Community in those treaties. In this respect there is no problem either with the UNFCCC or the Kyoto Protocol.

The UNFCCC contains a specific provision, namely Article 22, which explicitly foresees the possibility for 'regional economic integration organizations' to become Parties to the Convention together with its Member States. The same provision also states that in such a case, the regional economic integration organizations[86] and their Member States which are Parties to the Convention must 'decide on their respective responsibilities for the performance of their obligations under the Convention', and the regional economic integration organizations must 'declare the extent of their competence with respect to the matters governed by the Convention' in the framework of their instrument of ratification, acceptance, approval, or accession. This is essentially because the regional organizations and their Member States can in no case be entitled to exercise their rights under the Convention concurrently, as Article 22 UNFCCC carefully restates.

An almost identical provision to Article 22 UNFCCC is also contained in the Kyoto Protocol, in Article 24, where it is stated that 'this Protocol shall be open for signature and subject to ratification, acceptance or approval by States and regional economic integration organizations which are Parties to the Convention on climate change'. Article 24 Kyoto Protocol then contains specific rules on the assessment and the declaration of their respective responsibilities between the regional economic organizations and their Member States, which recall in their entirety the UNFCCC wording seen above.

Under the Kyoto Protocol, however, there is the possibility for some Parties, not necessarily those belonging to regional economic integration organizations, although this is the obvious case, to agree on the 'joint fulfilment' of their

---

[86] Despite the broad reference to the 'regional economic integration organization' as the recipient of this provision, in reality the only organization which can so far qualify for such a status is the European Community.

commitments pursuant to the provisions of Article 3 and 4 of the Protocol. The starting point for the 'joint fulfilment' of obligations agreed by the EC together with its Member States is Article 4 of the Kyoto Protocol, which prescribes that the Parties to it which have agreed to fulfil jointly their commitments 'shall notify the secretariat of the terms of the agreement on the date of deposit of their instruments of ratification, acceptance or approval of this Protocol, or accession thereto', including a specific reference to the 'respective emission level allocated to each of the Parties to the agreement'.

In accordance with this provision, the EC and its Member States, shortly after the signature of the Kyoto Protocol in June 1998, reached an agreement within the framework of the EC Council, on the respective level of emission limitation for each of the 15 Parties which were Members of the EC at that time. The EC 'joint fulfilment' of obligations based on Article 4 of the Kyoto Protocol has been called the 'EC Bubble' and so far represents the only concrete case of joint fulfilment of the Kyoto Protocol commitments, whereas the agreement reached by the EC and its Member States to give effect to the 'EC Bubble' is normally indicated as the 'Burden Sharing Agreement'.[87]

By means of the 'Burden Sharing Agreement', with a view to implementing the 'joint fulfilment of obligations' foreseen by the Kyoto Protocol, the common objective of an overall reduction of 8% of the combined level of GHG emissions (which is written in Annex B to the Kyoto Protocol both as the overall EC objective and the specific one for each of the 15 EC Members), was redistributed between the Member States. In particular, under the agreement reached, each Member State was allocated a differentiated contribution which was calculated by taking into account inter alia the specific expectations for economic growth, the energy mix, and the industrial structure of each Member State.

The EC Council, within the 1998 'Burden Sharing Agreement', further agreed that the terms of the agreement would be included in the Council Decision on the approval of the Protocol by the Community. Article 4(2) of the Protocol requires, in fact, the European Community and its Member States to notify the UNFCCC Secretariat with the terms of this agreement on the date of deposit of their instruments of ratification or approval.

The terms of the 'Burden Sharing Agreement' were therefore recalled and incorporated into the EC Council Decision 2002/358, the legal act by which the EC Council acceded to the Kyoto Protocol on behalf of the European Community.[88] Such a decision also addresses the issue of the demarcation of competence between the EC and its Member States, by including a Declaration on the respective competence regarding the implementation of the Kyoto Protocol in Annex III thereto.

---

[87] See Doc. 9702/98 of 19 June 1998 of the EC Council reflecting the outcome of proceedings of the Environment Council of 16–17 June 1998, Annex I.
[88] See EC Council Decision 2002/358, in OJ L130/2002.

The Declaration simply contains a very tautological restatement of the fact that in the climate change sector the EC and its Member States possess a shared competence, which has been and will continue to be exercised in a joint and coordinated way, as has been demonstrated by the relevant climate change legislation adopted so far. Unfortunately, however, the Declaration does not help in any way to better clarify for the third parties and Members to the Kyoto Protocol the exact boundaries of the EC and Member States' competence in the climate change sector. This, in turn, fails to shed more light on the related issues of their respective responsibilities in cases of non-fulfilment of their obligations under the Kyoto Protocol. This issue, however, will receive more in-depth analysis in next paragraph.

C. Issues of International and EC Responsibility for Non-compliance with the Climate Change Commitments under the Kyoto Protocol

The issue of the responsibility for non-compliance, particularly in the case of some Parties proceeding to the joint fulfilment of their obligations under the Protocol pursuant to the provision of Article 4, represents, in fact, a very interesting and specific characteristic of the Kyoto Protocol compliance regime. In such an instance, the issue of the responsibility of a Party for non-compliance may present different features depending on whether it is addressed from an international law or from an EC law perspective, as the following paragraphs will try to explain.[89]

In this respect, Article 4 of the Kyoto Protocol foresees the possibility for 'any Parties included in Annex I' of the UNFCCC to agree on the joint fulfilment of their obligations under the Kyoto Protocol, either outside or within the framework of a regional economic organization. This provision, obviously, has been created mainly to accommodate the needs of the European Community, which is so far the only non-State (*rectius*, regional economic integration organization) Party to the Kyoto Protocol together with its Member States, although it does not exclude the possibility for other Parties, which are not acting in the framework of such an organization, to proceed to the joint fulfilment of their obligations stemming from the Kyoto Protocol.

In this sense, it should be recalled first of all that once some of the Parties to the Kyoto Protocol have reached an agreement for the joint fulfilment of their obligations, and have duly notified it to the UNFCCC Secretariat, they become bound to their respective obligations written in the subsequent agreement, which essentially modifies their initial commitment originally written in Annex B to the Kyoto Protocol. However, as already mentioned above, the practical features of

---

[89] The present analysis is drawn extensively from Montini, supra n. 85, p. 106ff.; on this issue, from an EC law perspective see also Jacquemont, F., 'The Kyoto Compliance Regime, the European Bubble: Some Legal Consequences', in Bothe, M. and Rehbinder, E. (eds.), *Climate Change Policy* (Eleven International Publishing, 2005), p. 352ff.

their responsibility essentially depend on whether or not the Parties concerned, which have reached an agreement for the joint fulfilment of their obligations, are members of a regional economic integration organization.

In this sense, with respect to Parties acting outside the framework of a regional economic integration organization, Article 4(5) of the Kyoto Protocol states that:

> In the event of failure by the Parties to such an agreement to achieve their total combined level of emission reductions, each Party to that agreement shall be responsible for its own level of emissions set out in the agreement,

while in the case of Parties which belong to and act in the framework of a regional economic organization, such as the European Community, Article 4(6) of the Kyoto Protocol foresees that:

> If Parties acting jointly do so in the framework of, and together with, a regional economic integration organization which is itself a Party to this Protocol, each member State of that regional economic integration organization individually, and together with the regional economic integration organization acting in accordance with Article 24, shall, in the event of failure to achieve the total combined level of emission reductions, be responsible for its level of emissions as notified in accordance with this Article.

The two situations essentially differ depending on the involvement of the only regional economic integration organization which is so far a Party to the Kyoto Protocol, namely the European Community. In fact, in case one or more Member State(s) is (are) not fulfilling its (their) obligations up to an extent that its (their) failure also implies the failure of the EC to meet its overall collective target, the responsibility for the failure to fulfil their commitments will extend both to the defaulting State (or States) and to the European Community itself.

Several legal questions may arise in relation to such an issue. The first and foremost relates to the issue of the respective responsibilities of the Parties under international law, in case of a failure to satisfy the obligations contained in the joint fulfilment agreement. By reading the text of Article 4 to the Kyoto Protocol, it cannot, in fact, be easily determined whether in case of such a failure to fulfil their respective commitments both the EC and all its Member States will be responsible under international law or whether instead responsibility will arise just for the EC together with the defaulting State(s).

The wording of the Kyoto Protocol is not completely clear in this regard, and the question can only be solved by referring to the 'context' and the 'object and purpose' of the Kyoto Protocol, rather than sticking to the plain textual data, as mandated in such cases by the general rule on the interpretation of international treaties contained in Article 31 of the Vienna Convention on the Law of Treaties. In this context, in fact, if one places the provision of Article 4 of the Kyoto Protocol within the framework of its 'context' and its 'object and purpose', it emerges that the most logical interpretation can only be the one which makes the defaulting State(s) responsible towards the other Parties, together with the regional economic organization, namely the EC, rather than all the EC Member States which are Parties to the agreement on the joint fulfilment of their obligations.

Moreover, the second highly relevant question which arises here pertains to the additional responsibility which may derive under EC law from the failure of one or more of the EC Member States to fulfil their commitments under the joint fulfilment agreement. In such a situation, it is to be understood that besides the responsibility stemming from International law for the defaulting State or States (and possibly for the EC itself), there are some additional consequences which may arise under EC law for the relevant State(s) for failing to fulfil its (their) commitments under the Kyoto Protocol.

The main reason for this lies in the fact that the modified commitments undertaken by the EC Member States in the framework of the agreement on the joint fulfilment of their obligations, commonly indicated as the 'EC Burden Sharing Agreement', were agreed upon by the Environment Ministers of the EC Member States in 1998, but were then transferred into a binding Decision by the EC Council in 2002.[90]

This may obviously entail very relevant consequences under EC law, in a case where some of the commitments are not fulfilled by one or more of the interested Parties.[91] In such a case, a Party to the EC Burden Sharing Agreement that does not fulfil its obligations written into the agreement may be sanctioned under EC law on the basis of the combined application of Article 10 EC Treaty, which contains the general 'loyalty principle' that informs all the relationships of the EC Member States with the EC legal system and the EC institutions, as well as the specific provisions contained in Decision 2002/358/EC by means of which the EC Council approved the Kyoto Protocol on behalf of the European Community.

On the basis of such provisions, it can be assumed that an EC Member State which fails to fulfil its obligations under the Kyoto Protocol, as supplemented by the EC Burden Sharing Agreement, may be subject to an infringement procedure under Article 226 EC Treaty by the European Commission for its violation of the applicable EC law provisions just mentioned above.

In purely legal terms, this is the answer one can give to the question. In practical terms, however, it is still not clear what the concrete disadvantage of this double responsibility would be for a non-complying EC Member State, both under international law and under EC law. In fact, the effectiveness of the

---

[90] See EC Council Conclusions of 16–17 June 1998 (§ 2 and Annex I) and EC Council Decision 2002/358 of 25 April 2002 (Annex II) concerning the approval on behalf of the EC of the Kyoto Protocol.

[91] It should be recalled here that according to Article 4(4) Kyoto Protocol, 'If Parties acting jointly do so in the framework of, and together with, a regional economic integration organization, any alteration in the composition of the organization after adoption of this Protocol shall not affect existing commitments under this Protocol.' As a consequence, the EC Burden Sharing Agreement only relates to the 15 States which were Parties to the EC when the Kyoto Protocol was signed.

possible sanctions stemming from EC law, which may be much stricter and punitive in nature and therefore act as a greater deterrent for the States concerned, may be severely limited by the fact that the ascertainment of the position of a defaulting State under EC law with respect to the obligation in the Kyoto Protocol and the EC Burden Sharing Agreement would probably be issued very late in time, well after the expiry of the first commitment period foreseen by the Kyoto Protocol. This is essentially due to the fact that, nowadays, the usual duration of the infringement procedures performed under Article 226 EC Treaty is normally around two or three years. This is obviously not a very efficient duration, particularly in a sector such as that of climate change, which is characterized by high costs of compliance for the Parties and which should be assessed and planned as early as possible.

5. Concluding Remarks

In conclusion, it emerges from the analysis conducted above that the European Community is striving to become not simply a key international player in the environmental field, but rather a reference or benchmark model for most countries around the world as far as the development of their national environmental policy and law is concerned.

EC activism at the international level is supported by the relevant EC Treaty provisions, in particular Article 174(1) and 174(4), which stress the external environmental dimension of EC action as a key component of the EC's overall environmental policy and law. In fact, as we have already described above, the former provision inserts a reference to the promotion of *'measures at international level to deal with regional or worldwide environmental problems'* among the four basic objectives upon which the EC action in the environmental field ought to be based, while the latter explicitly recognizes a concurrent external competence of the European Community, which may imply the conclusion of international environmental treaties with third parties.

We have seen in the analysis provided above that the EC's proactive approach may sometimes cause some clashes with other existing international law provisions, such as for instance those on the free movement of trade in goods, or in any case may render the EC a forerunner which is not easily followed by most of the other players around the world in several other areas, such as in the climate change sector, for various reasons related to economic, social, environmental, and technological factors, as well as to lack of information and consciousness about the seriousness of the major environmental emergencies.

In both cases, a possible solution which should be carefully considered by the EC to try and impose its environmental view on other States around the world, without running the risk of breaching some of its international law obligations or of being isolated in a forerunning environmentally friendly approach, would probably consist in developing some (soft) ways to 'push' other States towards a (more or less) voluntary adoption of EC standards as a reference point for the development of their own national legislation.

An interesting embryonic example in this regard may come from the *Energy Community South East Europe Treaty* (ECSEE Treaty). This international agreement was signed in Athens (Greece) in October 2005 by the representatives of the European Community as well as by the trade and energy ministers of the following South East European countries: Albania, Bulgaria, Bosnia and Herzegovina, Croatia, the former Yugoslav Republic of Macedonia, Montenegro, Romania, Serbia, and the United Nations Mission in Kosovo (MIK—established pursuant to the UN Security Council resolution 1244).[92]

The ECSEE Treaty, which entered into force on 1 July 2006, aims at establishing a single and comprehensive regulatory framework for trading energy (electricity and gas) across South East Europe and the European Community. Its more interesting peculiarity, however, lies in the fact that it foresees a specific duty upon each of the Parties wishing to trade energy with the EC to implement within their national legislation, in compliance with specific timetables included in the annexes to the Treaty, the relevant *acquis communautaire* in the following four sectors: Energy, Environment, Competition, and Renewables.

With specific reference to the energy sector, the scope of application of the duty to implement the relevant *acquis communautaire* is represented by the so-called 'Network Energy', which includes the electricity and gas sectors falling within the scope of EC Directives 2003/54/EC and 2003/55/EC, establishing common rules of the EC's international market in the electricity and natural gas fields. Moreover, it is explicitly established by the ECSEE treaty that after its entry into force 'the construction and operation of new generating plants shall comply with the *acquis communautaire* on environment'.

Obviously, for the purpose of the ECSEE Treaty the '*acquis communautaire* on environment' refers only to some selected legal acts, which include the EC Directive 1985/337 on EIA (and subsequent modifications), EC Directive 1999/32 on the sulphur content of certain liquid fuels (and its subsequent amending Directive), EC Directive 2001/80 on the protection of the air from large combustion plants, and the specific provision of Article 4(2) of EC Directive 79/409 on the conservation of wild birds. In addition to the implementation of the above listed legal acts, the Parties to the Treaty 'shall endeavour' to accede to the Kyoto Protocol and to implement EC Directive 1996/61 on Integrated Pollution Prevention and Control (IPPC).

The ECSEE Treaty, therefore, represents a very good embryonic example for reflecting on what the European Community should carefully consider in order to be able to propose/impose the adoption and implementation of its environmental standards to third countries, not necessarily from the specific 'accession'

---

[92] See ECSEE Treaty, in <http://www.energy-community.org>.

perspective, but rather in the framework of broader cooperation agreements, possibly with a relevant economic dimension. This process, by which the EC could try to propose/impose to other States its 'environmental' standards, possibly linking their effective adoption and implementation to the provision of various other advantages for the States concerned in their relations with the EC itself, can be named 'environmental conditionality' and should be the object of a more careful examination, which is, however, beyond the scope of the present contribution.

6

A Climate of Change: An Analysis of Progress in EU and International Climate Change Policy

Jürgen Lefevere[1]

1. Introduction

Climate change policy and legislation has been one of, if not the, fastest moving areas of EU and international policy and law. The pace and number of international negotiations and meetings, combined with the large amount of new policy proposals, in particular in the EU, make this an exciting area to follow. But this also makes it a challenging issue to write about, without a contribution being out of date well before it is even published. It is for this reason that this chapter does not seek to give an overview of EU and international climate policy. Instead, it attempts to analyse, from an EU perspective, key events that took place during the year 2007 and in the first months of 2008—a time period that marks fundamental and hopefully sustained changes in both EU and international climate policy. In doing so, it will in particular look at some of the governance aspects of international climate policy, in line with the theme of this volume.

The fundamental changes in EU and international climate policy that took place during 2007 and the first months of 2008 were underpinned by a number of important new developments. Key among these was the adoption and publication of the reports of the three working groups of the Intergovernmental Panel on Climate Change (IPCC), providing an update of the latest state of climate science. All three working group reports, as well as the synthesis report itself, enjoyed unprecedented public attention. The attention for the work of the IPCC seamlessly built on the growing public discussion on climate change that received

---

[1] The views expressed in this contribution are personal and do not necessarily represent those of the European Commission. The author wishes to thank Benito Müller, Lars Müller, Lavanya Rajamani, and Joanne Scott for their comments on an earlier draft—all mistakes remain the responsibility of the author.

a significant boost from Al Gore's film 'An Inconvenient Truth' in mid-2006 and the publication of the Stern Review on the Economics of Climate Change in October 2006.[2] Al Gore's film, the Stern Review, and the IPCC's new reports not only dramatically increased the public awareness of the threat of climate change. They also increased the public's recognition of the impact of its own behaviour on global greenhouse gas (GHG) emissions and made a clear economic case for the need for international action. The granting of the Nobel Peace Prize in December 2007 jointly to Al Gore and the IPCC deservedly crowned their important contributions.

Fundamental changes also took place at the political level. The issue of climate change has for the past few decades often been seen as 'just' another environmental issue, rarely featuring in any serious manner in top level international politics. This changed dramatically during the course of 2007. This change was only partly driven by the increased public attention resulting from 'An Inconvenient Truth' and the IPCC reports. At least as important was the increasing awareness of the important opportunities offered, both economically and politically, by linking the issue of climate change to energy and energy security policy, but also competitiveness. This link was most apparent in the EU, where the Commission put forward, in January 2007, a proposal for a fully integrated EU climate change and energy strategy. This link was however also increasingly picked up in other countries, albeit to various degrees and for various reasons. The linking of climate change within particular energy, but also competitiveness issues and, more recently, broader sustainable development issues, has enabled climate change to enter the field of 'serious politics', providing it with the right level of attention needed to find a solution.

Both the previous two developments provided the foundation for the third key change during 2007: the mainstreaming of climate change in different areas of international politics. Until recently, international discussions on climate change were mostly limited to the negotiations within the meetings of the UN Framework Convention on Climate Change (UNFCCC) and its Kyoto Protocol. Over the past few years this has gradually changed. Significantly accelerating with the high-level attention for climate change under the United Kingdom's presidency of the G-8 in 2005, climate change has been increasingly mainstreamed within a number of bilateral and multilateral cooperation efforts. During 2007 this culminated in an impressive number of climate change related meetings across different UN fora, including the UN Security Council,[3] the UN General

---

[2] The full text of the Stern Review is available through: <http://www.hm-treasury.gov.uk/independent_reviews/stern_review_economics_climate_change/sternreview_index.cfm>.

[3] The security dimension of climate change was discussed at the 5663rd meeting of the UN Security Council on 17 April 2007. A report of the meeting is available through: <http://www.un.org/News/Press/docs/2007/sc9000.doc.htm>.

Assembly,[4] and a high-level meeting on climate change organized by the UN Secretary General Ban Ki-moon.[5]

Together, these three developments finally helped bring about the two major breakthroughs in international and EU climate change negotiations: the agreement on the Bali Action Plan in December 2007 and the proposals by the European Commission to implement the EU's climate change and renewable energy package in January 2008. Both recent events have not set the basis for the international negotiations that, if successful, will lead to the much needed global and comprehensive agreement on further international action on climate change in December 2009 in Copenhagen.

As signalled at the start of this introduction, this contribution will in particular look at some of the governance challenges following from the important new developments during 2007 and the first few months of 2008. This includes the rapid fragmentation of the work on climate change in terms of multiplicity of actors, processes, and institutions, the needs for mainstreaming, and also challenges for institutional organization. In order to do so, this contribution will start with an analysis of the three developments described above: the increased awareness of the science, the linking of climate change with broader energy and competitiveness politics, as well as the mainstreaming of climate change in the international policy agenda. It will then discuss the outcome of the UN climate change meeting in Bali and end with a short reflection on 'the way forward', analysing some of the challenges ahead. The EU climate change and energy package, proposed on 23 January 2008, is not discussed in this contribution—at the time of writing the proposal had only recently been put forward and discussions on this have only just started, making a short description less relevant for this contribution.

2. Climate Science: Spreading the Inconvenient Truth

A. Politics and IPCC's Fourth Assessment Report

In November 2007 the Intergovernmental Panel on Climate Change (IPCC) released its Fourth Assessment Report (AR4), providing an up-to-date and authoritative international scientific assessment of the state of climate science. The adoption of the report and the dates of the UN climate conference in Bali in December were deliberately scheduled to enable the IPCC's conclusions to be

---

[4] The UN General Assembly had an informal Thematic Debate on 'Climate Change as a Global Challenge' on 31 July and 1 August 2007. A report of the meeting is available through: <http://www.un.org/ga/president/61/follow-up/thematic-climate.shtml>.

[5] United Nations Secretary-General Ban Ki-moon held a high-level meeting, 'The Future in our Hands: Addressing the Leadership Challenge of Climate Change' at UN Headquarters in New York on 24 September 2007. More information on this meeting is available through: <http://www.un.org/climatechange/2007highlevel/>.

taken into account in the negotiations in Bali. The IPCC's AR4 was key in the important progress in the international climate change negotiations that was made during 2007. The important role that the IPCC's AR4 played in Bali was however by no means a given. Over the last five years, in particular since the rejection of the Kyoto Protocol by the Bush Administration, the IPCC had been subjected to various attempts to reduce the influence of its work.

The IPCC's scientists and chairs have a strong history of seeking to ensure that its work is balanced and neutral. This is particularly reflected in the way that the IPCC prepares its reports, where a good geographical spread of authors and an in-depth review of both the contributions and products of the IPCC are two of the key methods for achieving this. A short description of the way in which the IPCC works, reflecting this need for neutrality and comprehensiveness, is given in Box 1 below. The rigorous review process that IPCC assessment reports are subjected to, counters criticism from climate sceptics that IPCC reports are overly alarmist and do not reflect general scientific opinion.[6]

Box 1: The IPCC & the process for preparing its assessment reports

The IPCC was set up by the World Meteorological Organization (WMO) and by the United Nations Environment Programme (UNEP) in 1988.[7] Its role is 'to assess on a comprehensive, open and transparent basis the scientific, technical and socio-economic information relevant to understanding the scientific basis of risk of human-induced climate change, its potential impacts and options for adaptation and mitigation'.[8] The IPCC does not conduct its own research, but brings together the results of published research findings. The structure of the IPCC seeks to ensure objectivity and comprehensiveness through ensuring participation of experts from all regions in the world and a range of relevant disciplines and through a strict review process that includes both experts and governments.

The main products of the IPCC are its Assessment Reports. These reports, published on average every five to six years, provide a comprehensive update on the status of climate science. Almost every IPCC Assessment Report played a key role in the international climate change negotiations. The IPCC's First Assessment Report was released in 1990 and provided the basis for the UN General Assembly's Decision to

---

[6] Czech President Vaclav Klaus, a vocal climate sceptic, in his speech at the UN Secretary General's High Level Event on Climate Change stated that 'there is no scientific consensus about the causes of recent climate changes' and proposed that 'the UN should organize two parallel IPCCs and publish two competing reports. To get rid of the one-sided monopoly is a sine qua non for an efficient and rational debate.' See <http://www.un.org/webcast/climatechange/highlevel/2007/pdfs/czechrepublic-eng.pdf>.

[7] For an elaborate description of the history of the IPCC See IPCC Secretariat, '16 Years of Scientific Assessment in Support of the Climate Convention', available at <http://www.ipcc.ch/pdf/10th-anniversary/anniversary-brochure.pdf>.

[8] Paragraph 2 of the Principles Governing IPCC Work, approved at the IPCC's 14th session (Vienna, 1–3 October 1998), most recently amended at the 25th session (Mauritius, 26–28 April 2006), available at <http://www.ipcc.ch/pdf/ipcc-principles/ipcc-principles.pdf>.

initiate negotiations on the UNFCCC for completion prior to the UN Conference on Environment and Development in June 1992. The Second Assessment Report was completed in 1995 and provided important impetus for the Adoption of the Kyoto Protocol in 1997. The Third Assessment Report, released in 2001, came in a politically difficult climate, shortly after Bush's rejection of the Kyoto Protocol and at a time that the rest of the world was busy saving the Protocol in the run-up to the Marrakech climate change conference in November 2001.

A key strength of the IPCC's Assessment Reports is that they provide a complete overview of the state of climate science, not only limited to natural science. To ensure this broad view, contributions for IPCC's Assessment Reports are prepared through three Working Groups. Working Group 1 (WG 1) assesses the physical scientific aspects of the climate system and climate change. Working Group 2 (WG 2) assesses the vulnerability of socio-economic and natural systems to climate change, negative and positive consequences of climate change, and options for adapting to it. Working Group 3 (WG 3) assesses options for mitigating climate change through limiting or preventing GHG and enhancing activities that remove them from the atmosphere.

The draft report of each of the IPCC's working groups is prepared by a number of Coordinating Lead Authors and Lead Authors.[9] These authors are selected by the Bureau of the relevant Working Group, who should ensure appropriate representation of experts from developing countries, developed countries, and economies in transition, with at least one and normally two or more authors from developing countries. The Coordinating Lead Authors and Lead Authors may enlist other experts as Contributing Authors. The first draft of the report is prepared on the basis of contributions to the IPCC, peer reviewed and internationally available literature, as well as material unpublished or not peer reviewed that is subject to a separate IPCC assessment process, involving a review of the quality of the material and the validity of its sources.

Following the preparation of the first draft, the report is subject to a two-stage review process to ensure that the draft is balanced and complete. The review is coordinated by two review editors for each chapter that were not involved in the preparation or review of the material for which they are the editor. The first stage of the review is done through a circulation of the draft report to experts, with a request for comments. Following a revision, the new draft is then circulated to governments, to all the coordinating lead authors, lead authors, and contributing authors as well as expert reviewers, again with a request for comments. On the basis of these comments a final draft report is prepared by the coordinating lead authors and lead authors in consultation with the review editors. This final report is put forward for 'acceptance' to the relevant Working Group, meaning that it is not subject to line-by-line discussion and agreement but to a consensus that the report presents a balanced and objective view of the subject matter. Together with the draft final report the Working Group is also asked to 'approve' a summary for policy-makers. This is done through a detailed, line-by-line

---

[9] For a broader overview of the IPCC procedures see Procedures for the Preparation, Review, Acceptance, Adoption, Approval, and Publication of IPCC reports, Appendix A to the Principles Governing IPCC Work, n. 3 supra, available through <http://www.ipcc.ch/pdf/ipcc-principles/ipcc-principles-appendix-a.pdf>.

Box 1. *(cont)*

discussion and agreement, and which is also subject to an 'acceptance' by the Panel. In addition to the individual Working Group reports the IPCC also prepares a synthesis report, consisting of the main report, based on the three Working Group reports, which is put up for 'adoption' by the plenary, involving a process of endorsement section by section. The synthesis report also contains a summary for policy-makers which is subject to 'approval' through a line-by-line discussion and agreement.

It would however be wrong to argue that the IPCC's work is immune to political influence. The stage in which political manoeuvring is most obvious is the finalization and adoption of the IPCC's assessment report and the contributions of its working groups to those reports. Both the plenary of the Working Groups and the plenary of the IPCC as a whole consist of representatives of all Member Countries. The 'approval', 'adoption', and 'acceptance' of the different reports and summaries is usually done by consensus.[10] Only if consensus is deemed impossible can the IPCC decide to explain and record differing views, which in practice occurs only rarely. It should also be pointed out that the government representatives in IPCC meetings of a not insignificant number of IPCC Member Countries are the same as those in the climate negotiations and are often not necessarily scientific experts but rather negotiators. As a result, politicized debates are not uncommon during the negotiations on the summaries for policy-makers of the Working Groups and synthesis report. During the final stages of the negotiations on the various AR4 a number of countries were for instance adamant to ensure that the IPCC's work would remain neutral as to the different emission scenarios, in particular avoiding any preference for the lower emission scenarios advocated by the EU. Important examples of casualties of these negotiations include textual or graphical links between impacts of climate change and specific temperature increases. EU Member States advocated such examples to enable the use of the IPCC's AR4 to back up the EU's objective to limit global warming to 2°C above pre-industrial levels; other countries opposed such examples for precisely the opposite reason. At the same time countries sometimes seek to include conclusions on issues not addressed in the actual report. During the final stages of the adoption of the conclusions of the IPCC's WG3 for instance, India, China, and Brazil demanded a formal quantification of the historical responsibility of industrialized countries in their contribution to the build-up of greenhouse gases in the atmosphere.

Every time the IPCC releases an assessment report it is subject to sometimes very strong criticism from climate sceptics that these reports are alarmist and deliberately ignore scientific findings that are not in conformity with the view of

[10] See Box 1 for an explanation of these various terms.

the authors.[11] The thorough process for preparing the assessment reports however demonstrates that this criticism is unfounded. In fact the opposite conclusion could be drawn: the fact that these reports are adopted by consensus means that they may be more likely to understate the risks, identifying a 'lowest common denominator' that is acceptable to all, rather than overstating the risks of climate change.

Political pressure to reduce the impact of the IPCC's work was also obvious during the discussions on the scope, content, and timetable of the AR4, in particular also the question of whether and what kind of synthesis report would be prepared. For the previous IPCC assessment reports most of the public attention was drawn to the final product summarizing the full breadth of the report: the synthesis report and its summary for policy-makers. This synthesis report was prepared in addition to the reports of the three working groups and brought together the various results in a comprehensive analysis. During the decision-making process on the outline of the AR4, a number of countries had insisted that no synthesis report would be required at all, raising predominantly timing concerns, questioning its added value, and questioning its quality if it were to be prepared on too short a timetable. The compromise solution that was proposed by the IPCC chair was to limit the length of the synthesis report to 30 pages, with a summary for policy-makers of five pages.[12] The IPCC chair also proposed to contact the UNFCCC Executive Secretary with a request to postpone COP-13 by one month.[13]

A short summary of some of the key messages from the AR4's synthesis report, which was released in November 2007, is set out in Box 2 below. Initially it was feared that a short synthesis report would reduce the value of the AR4. The final AR4 synthesis report indeed attracted relatively little public attention. The main reason for this was however not the substantive conclusions of the synthesis report. Rather than focusing on this short report and the final outcome in November, the attention of the press had already been caught earlier in the process by the results of the discussions in the IPCC's three individual working groups. Working Group I released its report 'The Physical Science Basis' on 1 February, Working Group II its report 'Impacts, Adaptation and Vulnerability' on 5 April, and Working Group III its report 'Mitigation of Climate Change' on 3 May. The fact that the release of the contributions of the individual working groups was 'staggered' through the first half of 2007 greatly helped in sustaining the public attention that had been conveniently raised by Al Gore's film only a few months earlier and building up public awareness for the Bali UN climate negotiations, rather than focusing public attention on the release of the final report at the end of 2007.

---

[11] See for instance the comments by the Czech President Vaclav Klaus, n. 7 supra.
[12] See Earth Negotiations Bulleting, Summary of the 22nd session of the IPCC, 9–11 November 2004, n. 6 supra.
[13] n. 6 supra.

Box 2: Key messages from the IPCC's AR4[14]

a. Warming of the climate system is unequivocal: eleven of the last twelve years (1995–2006) rank among the twelve warmest years in the instrumental record of global surface temperature (since 1850). Between 1906–2005 the global average temperature has increased by 0.74°C.
b. Many natural systems are already being affected by regional climate changes and other effects of regional climate change on natural and human environments are emerging.
c. Global GHG emissions due to human activities have grown since pre-industrial times, with an increase of 70% between 1970 and 2004.
d. Most of the observed increase in globally-averaged temperatures since the mid-20th century is due to the observed increase in anthropogenic GHG concentrations. For the next two decades a warming of about 0.2°C per decade is projected.
e. The report gives a number of global and regional examples of impacts associated with different levels of global average temperature exchange, which includes for instance that by 2020 between 75 and 250 million people in Africa are projected to be exposed to increased water stress due to climate change, and agricultural production, including access to food, in many African countries is projected to be severely compromised.
f. Climate change is likely to lead to some irreversible impacts. For instance approximately 20–30% of species assessed so far are likely to be at increased risk of extinction if increases in global average warming exceed 1.5–2.5°C (relative to 1980–1999). This increases to 40–79% as global average temperature increase exceeds 3.5°C.
g. A wide array of options to adapt to the growing impact of climate change is available, but more extensive adaptation than is currently occurring is required to reduce vulnerability.
h. There is substantial economic potential for the mitigation of global GHG emissions over the coming decades that could offset the projected growth of global emissions or reduce emissions below current levels. An effective carbon-price signal could realize significant mitigation potential in all sectors. Delayed emission reductions significantly constrain the opportunities to achieve lower stabilization levels and increase the risk of more severe climate change impacts.
i. All stabilization levels assessed by the IPCC can be achieved by deployment of a portfolio of technologies that are either currently available or expected to be commercialized in coming decades.
j. In 2050, global average macro-economic costs for mitigation towards stabilization at the IPCC's lowest stabilization level (445ppm CO_2, roughly compatible with the EU's proposed targets) would lead to less that a 5.5% decrease of global GDP. This corresponds to slowing average annual global GDP growth by less than 0.12 percentage points.

[14] The full text of the Summary for Policy Makers, as well as the full report and its summaries, can be found at: <http://www.ipcc.ch>.

Political manoeuvring in the IPCC is not limited to the planning and adoption of the IPCC's reports, but also affects its organization. The most published incident involved the, for many unexpected, 'ousting' of the previous chair of the IPCC, the British-born US scientist Dr Robert Watson, in April 2002, and his replacement by Dr Rahendra Pachauri, the Indian Director General of The Energy and Resources Institute (TERI, formerly the Tata Energy Research Institute) in New Delhi. Dr Watson was a very vocal and proactive chair of the IPCC, whose role was increasingly seen as a threat by climate sceptics. Dr Watson's proactive role was one of the key reasons for the US choosing not to back his reappointment, but publicly supporting Dr. Pachauri in the run-up to the election of the new IPCC chair. Interestingly, shortly before the election of the IPCC chair the US environmental organization NRDC published a confidential memo from ExxonMobil to the White House. This memo was very critical of Dr. Watson's chairmanship of the IPCC and explicitly raised the issue of whether he could be replaced at the request of the United States, in addition to heavily criticizing specific scientists representing the US at the IPCC meetings.[15] This chain of events led to the accusation by inter alia environmental NGOs that the replacement of Dr. Watson by Dr. Pachauri was aimed at weakening the IPCC and reducing the impact of its work.

Under severe criticism from his moment of appointment, Dr. Pachauri was faced with a particular challenge to demonstrate his and the IPCC's scientific rigour and neutrality. Although a weakening of the IPCC's work may certainly have been the intention of some of those behind the events that led to the appointment of Dr. Pachauri, history has shown that this objective was not achieved. It would certainly not be appropriate to accredit Dr. Pachauri with the full responsibility for a truly collective effort such as the AR4, the publicity surrounding it, and in particular the 2007 Nobel Peace Prize for the IPCC, but his chairmanship did prove a key contribution to these achievements. What is perhaps an even more important, and for many unexpected, result of his chairmanship is the effect that the public attention for the AR4 and the IPCC's Nobel Prize has had on the awareness of climate change in developing countries, in particular India. Pachauri's developing country background has helped increase the acceptance and status of the work of the IPCC in developing countries. In India, the IPCC's Nobel Peace Prize has made Dr. Pachauri nothing short of a national hero, which in turn has had a very positive impact on the awareness of and public discussion on climate change in India.[16]

---

[15] The EXXON memo is available on the Internet at: <http://www.nrdc.org/media/docs/020403.pdf>.
[16] Dr. Pachauri was awarded the Padma Vibhushan, the second highest civilian honour in India, in January 2008. Earlier he was chosen 'Global Indian of the year' for the year 2007 by NDTV (New Delhi Television Limited), which is the largest Indian private television network.

B. Future Challenges for the IPCC

The last few years have shown the remarkable strength of the IPCC in maintaining its scientific integrity and strengthening its public profile. The fact that all of this took place in a period over which it had to cope with targeted efforts to undermine its work only underlines how much it deserved to be awarded the 2007 Nobel Peace Prize. The IPCC is now gearing up for its next challenges, central in which will be the possible preparation of a Fifth Assessment Report (AR5). In doing so, it will be faced with a number of new challenges. These include the participation of the EU in its further work, as well as the general future of the IPCC's structure and outputs. Both of these are briefly discussed below.

1. EU participation in the work of the IPCC

One of the challenges ahead for the work of the IPCC will be the way in which it will recognize the EU and its Member States. Although the European Community[17] is a Party to both the UNFCCC and its Kyoto Protocol (as a 'regional economic integration organization'), it does not have full participant status in the IPCC. The main reason for this is that Principle 7 of the Principles governing the IPCC work explicitly states that participation in the work of the IPCC is open to all UNEP and WMO member countries.[18] As the European Community is neither a member of UNEP nor of the WMO it has thus far not been given full participant status. In practice this means that the European Community, represented by European Commission officials, is only allowed to participate as an observer organization. As certain meetings during a Session of the Panel or a Working Group may be closed to observers, observer organizations are not admitted to any Session of the IPCC Bureau or Task Force Bureau and, most importantly, observers have no formal role in the approval of reports.[19] In addition to this, the absence of a role for the European Commission and the EU Presidency and the fact that little or no coordination of Member State positions takes place, means that in practice all EU Member States speak individually and on their own behalf at IPCC meetings.

The European Commission has been promoting a stronger coordination between Member States for their input in the IPCC process, focusing on non-scientific issues that frequently arise in IPCC discussions. These efforts have so far been met with scepticism. Member States fear a further politicization of the

---

[17] Under current EU law the European Union does not have legal personality, but the European Community does (Article 281 of the European Community Treaty). Membership of international organizations and the ratification of treaties is therefore in name of the European Community. The Treaty of Lisbon (OJ 17.12.2007, C306) however provides for the legal personality of the European Union, succeeding the European Community (new Article 46 A in the Treaty on European Union).

[18] Paragraph 7 of the Principles Governing IPCC Work, n. 3 supra.

[19] See the IPCC policy and process for admitting observer organizations, Adopted by the Panel at its 25th Session, 26–28 April 2006, available at <http://www.ipcc.ch/pdf/ipcc-principles/ipcc-principles-observer-org.pdf>.

IPCC's work, and see the participation of the European Community in addition to its Member States, as well as a coordination of EU positions, even on only political issues, as undermining the scientific independence of their representatives in the IPCC. In addition to that, and perhaps more importantly, Member States also fear a further loss of competence and freedom to contribute individually. On the other hand, the fact that the European Commission is a major funder of climate change related research and climate monitoring and observation, and a major financial contributor to the work of the IPCC, would merit enriching the debate in the IPCC with a European Community voice. Furthermore, the discussions in the UNFCCC have shown the important advantages following from a strong and unified EU contribution on political issues.

With the considerable and growing role of European Community funded climate research in the work of the IPCC, a European Community representative in the IPCC would seem warranted. In addition to that it is irrational that Europe's policy-makers speak with one voice in the climate negotiations, but European researchers and scientists are neither given a separate voice nor an opportunity to coordinate the EU's position on the many political issues discussed in the IPCC process. Fears that this would lead to a further politicization of the IPCC are unfounded and can be addressed by finding a modus operandi for the Community's participation that would guarantee the scientific independence of EU participants in the IPCC.

One way of ensuring this would be to change the IPCC Governing Principle number 7 to enable also Parties to the UNFCCC to participate in the work of the IPCC. The review of these Principles takes place at least every five years, the most recent review having taken place at the IPCC's 28th session in Budapest in April 2008, at which it was decided to continue the current Principles unchanged. The Budapest meeting did however discuss a proposal, to provide a special observer status to regional Economic Integration Organizations that are Parties to the UNFCCC and the Kyoto Protocol, but decided to postpone the discussion on this proposal to its next session, which is to take place in September 2008.[20]

2. *The future of the IPCC*

The discussion on the participation of the European Community in the work of the IPCC at the April meeting in Budapest came in the middle of a larger debate on the future of the IPCC.

---

[20] The European Community proposed that the IPCC adopt a decision stating that:
'The Panel decides that Regional Economic Integration Organisations (REIOs) that are Parties to the UNFCCC and the Kyoto Protocol take part in the IPCC works as "full participants". They are entitled to participate fully in the work of the Panel or any subsidiary body thereof. Such full participation includes the right to speak and the right of reply, as well as the right to introduce proposals and amendments, but not the right to vote, nor the right of being elected.' See for a summary of the Discussion: Earth Negotiations Bulleting, Summary of the 28th Session of the Intergovernmental Panel on Climate Change: 9–10 April 2008, available at <http://www.iisd.ca/download/pdf/enb12363e.pdf>.

In January 2008 the IPCC Secretariat distributed a discussion paper about the future of the IPCC and invited a broad range of actors involved in the work of the IPCC to provide their comments and views.[21] The seven-page discussion paper reviews the functioning of the IPCC in the past and analyses the need for change. It identifies four factors that may require such change:

- With better public awareness of climate change there is a demand for a higher level of policy relevance in the work of the IPCC.
- The analysis of the connection between climate change and sustainable development must be improved.
- There should be a greater emphasis on the economic aspects of climate change.
- There is a greater need for work on the regional aspects of climate change, in particular areas of the world for which such research is almost non-existent.

In order to respond to these challenges, the paper raises a number of issues for discussion, both in the organization of the IPCC's outputs and in the functioning of its Bureau. The main issue concerns the frequency and scope of future outputs of the IPCC. The paper states that suggestions have been made to favour a set of focused special reports rather than a comprehensive assessment like the AR4. It suggests that a set of special reports could be produced within the 5–6 year timeframe, with a comprehensive assessment then carried out every two cycles, although it explicitly states that a 10–12 year cycle for a comprehensive report would be too long.

Replacing the comprehensive report with a set of special reports would however bring a number of important disadvantages. A first disadvantage is that in a body with consensual decision-making like the IPCC it would be extremely difficult to come to an agreement on a list of topics for such reports and in particular also a prioritization in terms of timing for their preparation. These discussions could take up valuable time and could end up with a set of reports that reflects the full table of contents of the AR4. A further disadvantage is that a comprehensive IPCC report every five years has demonstrated its value in mobilizing public and political attention. Although in the case of the AR4 this attention was more focused on the reports of the working groups than on the synthesis report, a further splitting up of the work of the IPCC into smaller and more frequent reports could mean that the results of the IPCC's work lose their public impact.

The IPCC's April meeting in Budapest however followed the IPCC Chair's recommendation to continue the current system of a comprehensive assessment every five to six years. Despite some isolated calls for radical change, the vast majority of representatives were clearly in favour of continuing the current

---

[21] This paper, together with a synthesis of comments on this paper, is contained in document IPCC-XXVIII/Doc. 7, available through <http://www.ipcc.ch/meetings/session28/doc7.pdf>.

structure and working methods of the IPCC, with only minor evolutionary changes. The meeting thus decided to maintain the current working group structure. Importantly, it agreed to schedule the finalization of the WG 1 contribution to the 5th Assessment Report to be finalized by early 2013 and the contributions of WG 1 and 2 as well as the synthesis report as early as possible in 2014, which would allow the 5th Assessment Report to be finalized in that same year.

3. Climate Change, Energy Security, and Competitiveness

Although it has been on the international political agenda since the 1980s, climate change has, until recently, been ranked as 'just' an environmental issue. In fact if one examines the lists of participants from past UN climate change meetings, the vast majority of government representatives will come from environment ministries or (particularly in the case of a large number of developing countries) meteorological services. Although it would be wrong to argue that these do not have the mandate to represent their country or would not be sufficiently competent to participate in these negotiations, it can be concluded that these ministries or services are usually not ranked among the most powerful or politically influential. Being ranked as mostly an environmental issue meant that climate change suffered from the same 'trade-offs' that many environmental issues continue to suffer from today: either we protect our environment or we grow our economy. This trade-off has been most obvious in the climate policy of the Bush Administration, which on multiple occasions criticized environmental and climate change policies for their negative effects on the US economy.[22] The fact that the second preambular paragraph of the New Delhi Declaration, adopted at the 8th Conference of the Parties in New Delhi in November 2002, explicitly and up-front restates text from Article 4 paragraph 7 of the UNFCCC that states that 'economic and social development and poverty eradication are the first and overriding priorities of developing country Parties' demonstrates that developing countries too saw action to tackle greenhouse gas emissions as a threat to their economy, and many still continue to do so. India's presentation at the Vienna Climate Change talks in August 2007 for instance pointed at the tremendous costs of reducing its greenhouse gas emissions, money which it cannot afford to divert from its development activities.

This view has however become increasingly outdated over the past few years, for a number of reasons. Firstly, the emergence of a number of economic studies has shown that tackling climate change will actually have only a minimal impact or even a net benefit for the economy, especially when compared to the growing

---

[22] See for instance President Bush's first speech on climate change in June 2001, available at: <http://www.whitehouse.gov/news/releases/2001/06/20010611-2.html>.

economic costs of the adverse impacts of climate change. The much discussed Stern Review on the Economics of Climate Change for instance concluded that 'the overall costs and risks of climate change will be equivalent to losing at least 5% of global GDP each year, now and forever. If a wider range of risks and impacts is taken into account, the estimates of damage could rise to 20% of GDP or more. In contrast, the costs of action—reducing greenhouse gas emissions to avoid the worst impacts of climate change—can be limited to around 1% of global GDP each year.'[23] Similarly, the AR4 also underlined the low economic costs of global action on climate change, if action were to be taken early and using flexible instruments (Box 2 above). The EU's own analysis, done in preparation for the EU's climate change and energy package, discussed below, was very much in line with the general conclusions of the IPCC and the Stern Review, showing that investment in a low-carbon economy would require around 0.5 % of total global GDP over the period 2013–2030, reducing global GDP growth by only 0.19 % per year up to 2030, which is only a small part of the expected annual GDP growth rate of 2.8 %. This economic analysis came in addition to growing negative impacts of environmental pollution and climate change on the economic development of developing countries.[24] But perhaps decisive were the international concerns about energy security in a period of rapid price increases in the oil and gas market.

The increasing linking of climate change and other issues has been particularly visible in the EU. In early 2005, European Commission President Barroso still defended a need to focus on growth and jobs, with a priority over action on the environment.[25] This changed in the course of 2006. In a widely published article from 15 November 2006, President Barroso found that 'tackling climate change deserves its place at the top of the European Union's list of priorities'[26] and in his speech at the UN Secretary General's High Level Event on Climate Change in

---

[23] Stern Review, short executive summary, available at: <http://www.hm-treasury.gov.uk/media/9/9/CLOSED_SHORT_executive_summary.pdf>. Although Stern's figures have come under much criticism, both from economists and from countries, including India, the conclusion that mitigation will in the long run be cheaper than adaptation is generally supported by most other recent economic analysis.

[24] See for instance 'China must come clean about its poisonous environment', *Financial Times*, 3 July 2007 (<http://www.ft.com/cms/s/0/10f81a12-29a5-11dc-a530-000b5df10621.html>) and 'Pollution costs equal 10% of China's GDP' in the *Shanghai Daily* of 6 June 2006, quoting an official study of the Chinese State Environmental Protection Agency on the impact of local pollution in China (<http://www.chinadaily.com.cn/china/2006-06/06/content_609350.htm>).

[25] In his speech 'Working together for growth and jobs: a new start for the Lisbon Strategy' at the Conference of Presidents of the European Parliament in Brussels on 2 February 2005 he stated that 'Sustainable development remains the overarching goal that frames all our economic, social and environmental action. Let me say this. It is as if I have three children—the economy, our social agenda, and the environment. Like any modern father—if one of my children is sick, I am ready to drop everything and focus on him until he is back to health. That is normal and responsible. But that does not mean I love the others any less!' Available at <http://europa.eu/rapid/pressReleasesAction.do?reference=SPEECH/05/67&format=HTML&aged=0&language=fr&guiLanguage=en>.

[26] José Manuel Barroso, 'Climate change and energy: Europe's determination', also available at <http://ec.europa.eu/commission_barroso/president/pdf/article_20061115_en.pdf>.

September 2007,[27] President Barroso stated that 'a low carbon economy will be a stimulus to our mutual prosperity, not a brake on growth. Using energy more efficiently means saving money. Switching to cleaner energy sources improves our air quality and our health. Investing and innovating creates industrial know-how and sustainable jobs. Investing in renewable energy strengthens the security of our energy supplies.'[28]

The European Commission's new interest in climate change as part of an energy security and economic growth agenda was caused by a number of factors. A first and crucial factor was Europe's shocking confrontation with the vulnerability of its energy supplies in January 2006. Between 1 and 4 January 2006 Russia cut gas exports to the Ukraine following Ukraine's refusal to accept a hefty increase in natural gas prices, aligning what was previously a preferential rate with international market prices.

On 1 January 2007 Russia stopped the gas export to Azerbaijan and, more worrying for the EU, between 8 and 10 January interrupted oil exports through the Druzhba pipeline which runs through Belarus, following a dispute with Belarus on gas prices. This pipeline carries oil from south-east Russia to points in Ukraine, Hungary, Poland, and Germany and is the largest principal artery for the transportation of Russian (and Kazakh) oil across Europe. Although strongly denied by Russia, both events raised a strong suspicion in Europe that Russia could use its position as the EU's dominant energy supplier for political goals. These events came together with a realization that Europe is becoming increasingly dependent on imported hydrocarbons. Assuming a 'business as usual' scenario, the EU's energy import dependence will jump from 50% of total EU energy consumption today to 65% in 2030. Even more alarming is that the reliance on imports of gas is expected to increase from 57% to 84% by 2030, and of oil from 82% to 93%.[29] Following these events, new tools to limit the EU's energy consumption and increase the 'domestic' supply of (mostly renewable) energy significantly increased in popularity.

A second crucial factor for the EU's growing interest in an integrated climate change, energy security and economic growth agenda was the fate of the European Constitution. The EU suffered a severe identity crisis after the French and Dutch electorates rejected the European Constitution in late May and early June 2005. The difficult and protracted discussions on its follow-up necessitated a new champion cause to revitalize the European debate among the EU's citizens. Various public surveys showed not only the growing public interest in climate

---

[27] n. 5 supra.
[28] This speech is available at <http://europa.eu/rapid/pressReleasesAction.do?reference=SPEECH/07/563&format=HTML&aged=0&language=EN&guiLanguage=en>.
[29] Commission Communication, An energy policy for Europe, COM(2007)1 of 10 January 2007, available at <http://eur-lex.europa.eu/LexUriServ/LexUriServ.do?uri=COM:2007:0001:FIN:EN:PDF> and European Commission and Eurostat: Energy and Transport in Figures, Statistical Pocketbook 2007, available at <http://ec.europa.eu/dgs/energy_transport/figures/pocketbook/2007_en.htm>.

change, but, most importantly, also a clear support for strong action on both energy and climate change issues at an EU level.[30] Similar support was also increasingly visible at the political level. In the run-up to the Lahti informal European Council meeting in October 2006, the Dutch and UK prime ministers sent a joint letter to the Finnish Presidency, calling for a linking of the energy security, climate change, and economic growth debates and strong action at EU level on these issues,[31] a call that was supported by the Council. Conveniently, the linking of climate change and energy issues also provided a stronger foundation for EU action on energy policy, as the European Community Treaty does not provide a separate legal basis for a common energy policy, requiring this policy to be introduced through the Treaty's legal basis on competition, the internal market, or the environment.

A third factor was the increasing realization that clean technology and domestically generated energy also create jobs within Europe. The Commission's implementation package of the EU climate change and energy strategy, proposed on 23 January 2008, makes this very explicit.[32] Commission estimates underpinning that package show that renewable energy technologies already account for a turnover of €20 billion and have created 300,000 jobs within the EU. The EU's target of a 20% share for renewables is estimated to mean almost a million jobs in this industry by 2020—more if Europe exploits its full potential to be a world leader in this field. In addition, the renewable energy sector is labour intensive and reliant on many small and medium-sized enterprises, spreading jobs and development across Europe, and the same is true of energy efficiency in buildings and products. The Commission also estimates that the 'ecoindustry' already accounts for some 3.4 million jobs in Europe and is a growing industry that now accounts for over €227 billion in annual turnover. With a growing global interest in energy efficient and clean technologies, the EU has a real interest in ensuring that it strengthens its position among the first entrants into this market.

In the context of these developments and with the outcome of the Stern Review and the Commission's own analysis of the impact of climate change and

---

[30] EU citizens' growing support for EU rather than national action on both energy and environment policies has been going hand in hand. In late 2004, Citizens perceive both the European Union (33%) and national governments (33%) as the best levels for environmental decision-making. In late 2007 67% of European citizens preferred environmental decisions to be made jointly within the EU. In the field of energy policy, a Eurobarometer survey conducted in late 2005 showed that 47% of EU citizens believe that Europe is the best level for determining energy challenges (compared to 37% in favour of the national level). A Eurobarometer conducted in late 2006 showed that that this majority had grown to 62% (compared to 32% favouring the national level). The various Eurobarometer reports are available at <http://www.ec.europa.eu/public_opinion/archives/eb_special_en.htm> and <http://ec.europa.eu/public_opinion/archives/flash_arch_en.htm>.
[31] Letter by Dutch Prime Minister Balkenende and UK Prime Minister Blair to the Finnish Prime Minister Vanhanen of 19 October 2006, available at <http://www.minaz.nl/dsc?c=getobject&s=obj&objectid=92057>.
[32] Commission Communication, 20 20 by 2020: Europe's climate change opportunity, Brussels, 23 January 2008, COM(2008)30, available at <http://ec.europa.eu/commission_barroso/president/pdf/COM2008_030_en.pdf>.

the costs of mitigation action, the logical conclusion to link energy, climate change, and competitiveness as part of an integrated package was very quickly drawn. The European Commission's Proposals for an integrated climate change and energy package of 10 January 2007 explicitly took these three pillars as their basis.[33] The strength of this approach was confirmed when, in March 2007, the European Council under firm German leadership, followed the Commission's initiative and adopted the first EU climate change and energy strategy. A short summary of this strategy is given in Box 3 below.

Box 3: The EU's March 2007 climate change and energy strategy

---

The integrated climate change and energy strategy adopted by the European Council on 9 March 2007[34] was closely based on the European Commission's proposals from 10 January 2007.[35] Key elements of this strategy are:
- Confirmation of the strategic objective of limiting the global average temperature increase to not more than 2°C above pre-industrial levels.
- To launch negotiations on a global and comprehensive post-2012 agreement at the UN international climate conference at the end of 2007 and to complete these negotiations by 2009.
- Absolute emission reductions are the backbone of a global carbon market. Developed countries should continue to take the lead by committing to collectively reducing their GHG emissions in the order of 30% by 2020 compared to 1990 and by 60% to 80% by 2050.
- An EU objective of a 30% reduction in GHG emissions by 2020 compared to 1990 as its contribution to a global and comprehensive agreement for the period beyond 2012, provided that other developed countries commit themselves to comparable emission reductions and economically more advanced developing countries to contributing adequately to their responsibilities and respective capabilities.
- Until a global and comprehensive post-2012 agreement is concluded, and without prejudice to its position in international negotiations, the EU makes a firm independent commitment to achieve at least a 20% reduction of GHG emissions by 2020 compared to 1990.

---

[33] See Commission Press Release, Commission proposes an integrated energy and climate change package to cut emissions for the 21st Century, 10 January 2007, available at <http://europa.eu/rapid/pressReleasesAction.do?reference=IP/07/29&format=HTML&aged=0&language=EN&guiLanguage=en>; Commission Communication, An Energy Policy for Europe, n. 30 supra; and Commission Communication, Limiting Global Climate Change to 2 degrees Celsius: The way ahead for 2020 and beyond, COM(2007)2 of 10 January 2007, available through <http://eur-lex.europa.eu/LexUriServ/LexUriServ.do?uri=COM:2007:0002:FIN:EN:PDF>; as well as a range of supporting documents, available through <http://ec.europa.eu/energy/energy_policy/documents_en.htm> and <http://ec.europa.eu/environment/climat/future_action.htm>.
[34] Presidency Conclusions of the Brussels European Council, 8–9 March 2007, available at: <http://www.consilium.europa.eu/ueDocs/cms_Data/docs/pressData/en/ec/93135.pdf>.
[35] n. 34 supra.

Box 3. *(cont)*

- The need for developing countries to address the increase in these emissions by reducing the GHG emission intensity of their economic development, in line with the general principle of common but differentiated responsibilities and respective capabilities.
- Continue and further strengthen support for developing countries in lessening their vulnerability and adapting to climate change.
- Measures to strengthen the internal market for gas and electricity.
- Measures to strengthen EU solidarity on the security of supply, notably in the event of an energy supply crisis.
- Measures to increase energy efficiency in the EU, saving 20% of the EU's energy consumption compared to projections for 2020.
- A binding target of a 20% share of renewable energies in overall EU energy consumption by 2020.
- A 10% binding minimum target to be achieved by all Member States for the share of biofuels in overall EU transport petrol and diesel consumption by 2020.
- Request the Commission to table a European Strategic Energy Technology Plan.
- Member States and the Commission to work towards strengthening R & D and developing the necessary technical, economic and regulatory framework to bring environmentally safe carbon capture and sequestration (CCS) to deployment with new fossil fuel power plants, if possible by 2020.
- Commission to establish a mechanism to stimulate the construction and operation by 2015 of up to 12 demonstration plants of sustainable fossil fuel technologies in commercial power generation.

The EU is however not the only region for which the integration of energy security, climate change, and economic development or competitiveness issues has increased in importance. Higher energy prices and increasing competition for energy resources also led other countries to realize the joint opportunities of tackling climate change and energy security but also for tackling local air pollution. As a result of this, developing countries too are increasingly taking domestic action that reduces the growth of their greenhouse gas emissions. These actions are however still predominantly taken for reaching different objectives. China's 11th 5-year plan, endorsed by the People's Congress in March 2006, for instance set the objective to increase China's energy efficiency per unit of GDP by 20% by the year 2010, at 4% per year. In 2005 China had also set a renewable energy target of 15% by 2020, a doubling compared to its 2005 level. Both objectives primarily seek to increase China's energy security, but are compatible with climate change objectives. Similarly India has set energy efficiency and renewable energy targets (10% of added electric power capacity during 2003–2012 is to be sourced from renewable energy), not only for energy security reasons, but also to support its goal to provide universal access to electricity for its population.

Although developing countries are increasingly seeing the co-benefits of tackling climate change and the opportunities to use climate related policies to support other policy objectives, many of them remain reluctant to place such policies explicitly in the context of climate change. The reasons for this include a reluctance to be seen to take autonomous action on climate change, without stronger commitments for technological and financial support from developed countries, as well as a continued insistence by key developing countries that climate change is a problem created by the industrialized world, which should therefore also take the first steps to tackle it. There is nonetheless a growing understanding that tackling climate change is a crucial part of ensuring sustainable economic growth.

The reluctance to take concrete and sufficiently ambitious steps to realize co-benefits is not restricted to developing countries alone. Also in developed countries there remains a considerable gap between the political discussions on co-benefits and taking sufficiently ambitious and effective action that will lead to the GHG emission reductions that are required to avoid dangerous climate change. The steps taken in the EU are among the most progressive for doing so, but they will need to be followed up by concrete actions to demonstrate the EU's commitment and conviction that it can strengthen its economy by tackling these challenges jointly. The package, proposed by the European Commission on 23 January 2008, is a further step in that direction, as is the call of the European Council of March 2008 to reach agreement on the contents of the package by the end of 2008, enabling its adoption as early as possible in 2009.[36] The EU now faces the challenge of demonstrating that its conviction is deep enough to weather a slowdown in economic growth that the EU may be facing following the now global credit crisis.

4. Mainstreaming Climate Change into the International Policy Agenda

As already stated in the previous paragraph, climate change was until a few years ago mostly seen as an environmental issue and thus discussed among mostly environment ministers and meteorologists. This also meant that climate change discussions remained relatively 'isolated' within meetings under the UNFCCC and its Kyoto Protocol and rarely featured on the agenda of other international

---

[36] See paragraphs 17 to 29 (in particular paragraph 18) of the Presidency conclusions of the European Council of 13–14 March 2008, available at <http://www.eu2008.si/en/News_and_Documents/Council_Conclusions/March/0314ECpresidency_conclusions.pdf>.

fora or bilateral meetings.[37] Only over the past few years has the issue managed to penetrate high-level international politics. This has not only been visible within the EU, where climate change now regularly features on the agenda of EU Heads of State and Government, meeting in the European Council, but also internationally.

A. The G-8: from Gleneagles via Heiligendamm to Hokkaido

The United Kingdom played a key role in raising the issue of climate change on the international policy agenda, but also in introducing it in bilateral and multilateral discussions outside those under the UNFCCC. This important achievement was helped by the combination of the UK's presidency of the G-8 during 2005 and its presidency of the EU during the second half of 2005.

The UK's first major achievement was placing climate change and energy firmly on the agenda of the G-8 meeting in Gleneagles in July 2005. Although the Gleneagles outcome was handicapped by fierce US opposition to EU proposals on post-2012 action and the terrorist attacks in London (which required Prime Minister Blair to leave for London before the issue of climate change had been discussed), its outcome was significant in a number of ways.

One of the main deliverables of the Gleneagles Summit was the agreement on the Gleneagles Plan of Action.[38] The Gleneagles Plan of Action contains a series of relatively concrete actions to be undertaken jointly by the G-8 members. This includes actions on 'transforming the way we use energy', including for improving the energy efficiency of buildings and appliances, and the promotion of cleaner transport and industry, actions for 'powering a cleaner future', including cleaner fossil fuels and renewable energy, 'promoting networks for research and development', 'financing the transition to cleaner energy', 'managing the impacts of climate change', and 'tackling illegal logging'. Perhaps one of the most important results of the Gleneagles Plan of Action is however the fact that it directly engaged the International Energy Agency (IEA) and the World Bank and other International Financial Institutions (IFIs). Not only did it ask them to assist with its implementation, it also greatly helped in raising the issue of climate change on the agenda of these institutions. The IEA was asked to develop and provide advice on alternative energy scenarios and strategies aimed at a clean energy future. As a result of this request, in June 2006 the IEA published the 'Energy Technology Perspectives 2006'. Their contribution to the implementation of the Gleneagles Plan of Action also includes further work on energy efficiency in buildings, appliances, transport, and industry as well as a

---

[37] The actual adoption of the UNFCCC and its Kyoto Protocol, but also the intensive diplomatic efforts to keep the Kyoto Protocol alive following its rejection by the US were important exceptions to this.
[38] See for all documents and summit outcomes of the UK G-8 Presidency: <http://www.g8.gov.uk>.

strengthened international cooperation through various networks and initiatives.[39] The World Bank and other international financial institutions were asked to create a new investment framework for clean energy and development, including investment and financing,[40] setting the stage for the growing discussions on financing and investment for tackling climate change and its impacts. The engagement of both the IEA and the World Bank has been key in raising the profile of climate change and mainstreaming it in the energy and financing political communities.

Secondly, the Gleneagles Summit started the 'Gleneagles Dialogue', a regular dialogue on innovative ideas and new measures to tackle climate change among environment and energy ministers of the 20 countries with the greatest energy needs. This group includes the G-8 members as well as the '+5' emerging economies: Brazil, China, India, Mexico, and South Africa.[41] The Gleneagles Dialogue also monitors the implementation of the Gleneagles Plan of Action and as such provided an important forum for the discussion and follow-up to, in particular, the work of the World Bank and the IEA.

The UK's second major achievement during 2005 was its work with the European Commission to place climate change high on the agenda of EU summits with key third countries. This resulted inter alia in the EU-China Partnership on Climate Change, agreed at the EU-China Summit in Beijing on 2 September, which set the stage for an important intensification of the dialogue and cooperation between China and the EU on climate change,[42] including the work on a Carbon Capture and Storage demonstration plant in China.[43] In addition to this, the EU-India Summit on 7 September agreed to launch an EU-India Initiative on Clean Development and Climate Change, as part of the Joint Action Plan under the India-EU Strategic Partnership.[44]

These two key results[45] of the UK's combined G-8 and EU Presidencies not only helped place climate change squarely on the G-8 agenda, but also strengthened the role of climate change in the EU's bilateral relationships with third countries.

The Gleneagles outcome on climate change in particular provided a good basis for Germany's ground-breaking achievements during the Heiligendamm summit

---

[39] For an overview of the IEA's work following the Gleneagles Summit, see <http://www.iea.org/G8/index.asp>.

[40] See <http://siteresources.worldbank.org/EXTSDNETWORK/Resources/2007_CleanEnergyFrameworkBooklet_Final.pdf?resourceurlname=2007_CleanEnergyFrameworkBooklet_Final.pdf>.

[41] For more information on the Gleneagles Dialogue and its outcomes, see <http://www.defra.gov.uk/environment/climatechange/internat/g8/index.htm>.

[42] See <http://europa.eu/rapid/pressReleasesAction.do?reference=MEMO/05/298&format=HTML&aged=1&language=EN&guiLanguage=en>.

[43] See <http://www.defra.gov.uk/environment/climatechange/internat/devcountry/china.htm>.

[44] See <http://ec.europa.eu/external_relations/india/sum09_05/05_jap_060905.pdf>.

[45] These achievements of course come in addition to the UK's key role as EU Presidency during the UN Climate Change Negotiations in Montreal in December 2005, which for the first time opened an (informal) dialogue on the future international action on climate change.

in June 2007.[46] Strengthened by the agreement by the EU Heads of State and Government on the EU's climate change and energy strategy in March 2007, Germany negotiated hard, against fierce US opposition, for an ambitious outcome on climate change in Heiligendamm. An agreement was only found after a personal discussion between Merkel and Bush on the Wednesday afternoon before the formal start of the G-8 meeting. As a result of this agreement, the Heiligendamm Summit Declaration on 'Growth and Responsibility in the World Economy' contains not only a commitment to 'consider seriously the decisions made by the European Union, Canada and Japan which include at least a halving of global emissions by 2050', but also a 'call on all parties to actively and constructively participate in the UN Climate Change Conference in Indonesia in December 2007 with a view to achieving a comprehensive post-2012 agreement (post-Kyoto agreement) that should include all major emitters'.[47] Both hard-fought elements of the declaration provided an important impetus for the UN climate negotiations in Bali in December 2007. They also provided a good starting point for the Japanese G-8 presidency during 2008. Japan managed to secure agreement among G-8 leaders to 'seek to share with all Parties to the UNFCCC the vision of, and together with them to consider and adopt in the UNFCCC negotiations, the goal of achieving at least 50% reduction of global emissions by 2050' and to 'implement ambitious economy-wide mid-term goals in order to achieve absolute emissions reductions'.[48]

B. The Major Economies Meeting

Germany's ambitious proposals for the G-8 summit and its willingness to push these until the last minute raised the political pressure on the US to such a level that it had to respond at the last minute with its own counter-initiative: its initiative for a major emitters' process.

The idea for a major emitters' process was only put forward by the US a week before the Heiligendamm summit. The US, under pressure from other G-8 members and the general public for blocking progress on climate change in preparation for Heiligendamm, proposed a parallel negotiating process among the fifteen top emitters and energy consumers, with the intention to create a 'post-Kyoto'[49] framework by 2008. The objective of such process would be to commit

---

[46] For background documents on the German G-8 Presidency and the Summit, See <http://www.g-8.de>.
[47] See paragraphs 49 and 52 of the Summit Declaration, available at: <http://www.g-8.de/Content/EN/Artikel/__g8-summit/anlagen/2007-06-07-gipfeldokument-wirtschaft-eng,templateId=raw,property=publicationFile.pdf/2007-06-07-gipfeldokument-wirtschaft-eng>.
[48] See the 'G8 Hokkaido Toyako G-8 Summit Leaders Declaration', Hokkaido, Toyako of 8 July 2008, available at: <http://www.g8summit.go.jp/eng/doc/doc080714_en.html>.
[49] The term 'post-Kyoto' was introduced by the US and other Kyoto-sceptics to underline the need to replace Kyoto with another treaty. As the Kyoto Protocol does not end in 2013 and all Kyoto Parties are already since the Montreal UN climate meeting in 2005 engaged in negotiating further targets for developed countries under the Kyoto Protocol, this term is factually incorrect.

to a long-term non-binding GHG emissions goal and allow each country to establish its own mid-term national GHG emissions or intensity goal. The process would also pick up on the work initiated under the Asia Pacific Partnership and look at sectoral approaches for specific energy intensive industries and support parallel national commitments to promote specific clean energy technologies. Finally, the process would aim to create a review mechanism to ensure national progress towards the global goal. The US offered to host a meeting of top emitters and energy consumers in the second half of 2007.[50]

The US proposals were strongly opposed by the majority of other G-8 partners, in particular those from the EU. The objective of the process proposed by the US (a non-binding long-term goal and countries setting their own national objectives) was far from the ambitions that the EU had set itself for an international climate change agreement. Most importantly, it was feared that the creation of a new parallel process among a limited group of major emitters would severely undermine the UN climate treaties, the negotiations on a post-2012 agreement and the Bali meeting in December 2007.

The final text of the Heiligendamm statement contains a very much watered down version of the US proposal. It refers to the need to continue to work with major energy consuming and greenhouse gas emitting countries and welcomes the offer by the US to host a meeting of those countries in 2007. It also sets out some of the issues to be discussed among major emitters, including 'inter alia, national, regional and international polices, targets and plans, in line with national circumstances, an ambitious work programme within the UNFCCC, and the development and deployment of climate-friendly technology'. Importantly, the text also explicitly states that 'the dialogue will support the UN climate process and report back to the UNFCCC'.[51] In the run-up to the US meeting, the major economies process was renamed the Major Economies Meeting (the MEM), following pressure from China and in particular India not to apply the term 'major emitters' to them, and from the EU who did not recognize this as a process, but one or several meetings feeding into the UN climate change process.

The MEM meetings have de facto been the first opportunity, since the rejection of the Kyoto Protocol by President Bush in 2001, to (re-)engage the US in international discussions on a post-2012 agreement. It was interesting that the first meeting of the MEM was organized on 27–28 September in Washington, which proved to be a self-constructed Trojan horse, bringing high-level delegations, many side-events, and much press attention on climate change to the US capital. Expectations for the outcome of the MEM meetings were however relatively low in view of the continued reluctance of the Bush administration to put forward concrete proposals to mitigate its greenhouse gas emissions. The

---

[50] The original proposal from 31 May 2007 can be found at <http://www.state.gov/g/oes/rls/fs/2007/92156.htm>.
[51] Heiligendamm Summit Declaration, n. 48 supra, para. 53.

MEM did however come very close to what could have been a groundbreaking agreement on a 2050 global reduction objective and an interim objective for 2020. The last minute lack of US support for these objectives however significantly reduced the outcome of the MEM summit that was held in the context of the G-8 meeting in Hokkaido.[52]

C. Mainstreaming Climate Change in the UN System

During 2007 we've also observed a growing interest in climate change in the broader UN system. The new UN Secretary General Ban Ki-moon in early 2007 indicated that he intended to organize a summit on climate change, in support of the UN climate negotiations. In May 2007 he appointed three Special Envoys for Climate Change (the Norwegian ex-Prime Minister Gro Harlem Brundtland, Former President of Chile Ricardo Lagos Escobar, and Han Seung-soo, the former General Assembly President). The first High Level Event on Climate Change 'The Future in our Hands: Addressing the Leadership Challenge of Climate Change' was held on 24 September in New York and was attended by a large number of Heads of State and Government.[53] Although the event provided mostly a platform for statements on a range of climate related issues and did not see actual negotiations on further global action on climate change, it did play a key role in sustaining political and public attention in the run-up to the Bali meeting.

The UN Secretary General's meeting was preceded by a UN General Assembly (UNGA) informal thematic debate on 'climate change as a global challenge' from 31 July to 2 August 2007.[54] This event, originally scheduled for two days, had to be extended by a day to accommodate the number of speakers interested in giving a statement. Like the September High Level Event, it did not have a concrete outcome. It was however important in strengthening the awareness of climate change in other parts of the UN system and helping to build the momentum towards the Bali meeting. The UNGA debate was recently followed by a second thematic debate on 11 and 12 February 2008, entitled 'Addressing Climate Change: The United Nations and the World at Work'.[55]

These two UN meetings were certainly not the only climate related meetings organized within the UN system during 2007, but they clearly indicate the growing interest in the issue throughout the UN system. One of the interesting

---

[52] See the 'Declaration of Leaders Meeting of Major Economies on Energy Security and Climate Change', in Toyako, Hokkaido, Japan, on 9 July 2008, available at: <http://g8summit.go.jp/eng/doc/doc08709_10_en.html>.

[53] The meeting was attended by top officials from over 150 nations, including 80 heads of State or Government. See n. 5 supra.

[54] For a summary of this event see Chris Spence, Earth Negotiations Bulletin, Summary of the Informal Thematic Debate of the UN General Assembly on Climate Change as a Global Challenge, available at <http://www.iisd.ca/climate/unga/UNGA%20Climate%20Change%20briefing%20note.pdf>.

[55] See <http://www.un.org/ga/president/62/ThematicDebates/themclimatechange.shtml>.

concrete results of these various events was that the UN in August 2007 set up an Internet Gateway to the UN System's work on climate change, to better inform its various bodies and the general public about the different work streams on climate change.[56] This new Internet portal is however only one, albeit visible, result of a greater effort within the UN system to better coordinate its work on climate change. Work on this had started in early 2007, following a meeting of the UN's Chief Executives Board.[57] The need for this work was further underlined by UNGA Resolution 62/8 of 19 November 2007, in which the UNGA requested the UN Secretary General to submit, by 25 February 2008, a comprehensive report providing an overview of the activities of the UN system in relation to climate change.[58] A first discussion of the various papers was held in the context of the Bali meeting and the UNGA thematic debate in February. Various options for strengthening the ability of the UN system to deliver as one in relation to climate change are currently under consideration.[59]

D. Finance and Trade Ministers' Meetings in Bali

During the inter-sessional UN climate negotiations in Vienna in August 2007, the upcoming Indonesian Presidency of the UN climate negotiations in Bali in December 2007 surprised Parties with its announcement that it was intending to organize a meeting of finance ministers and a meeting of trade ministers in the context of the Bali meeting. The informal trade ministers' dialogue on climate change was held on 8 and 9 December in Bali, followed by a finance ministers' dialogue on 11 December. At the finance ministers' meeting there was much discussion on the role of the global carbon market and the need to integrate climate change into development policies. The trade ministers' meeting discussed more broadly the links between international trade and climate change and ways to strengthen the dialogue on this link.[60] In view of the busy schedule of meetings in the second half of 2008 and the short time there was to prepare these meetings,

---

[56] See <http://www.un.org/climatechange>.
[57] See the Summary of the Discussions of the UN System Chief Executives Board fall 2007 Session Retreat, <http://hlcp.unsystemceb.org/documents/HLCP15th/CEB2007-2-Retreat-Summary%208Feb.pdf>, as well as the paper 'Co-ordinated UN System Action on Climate Change', 22 November 2007, available at: <http://www.uneca.org/eca_programmes/sdd/events/climate/CEB-Climate.pdf>.
[58] UNGA, 62nd session, Resolution 62/8, Overview of United Nations activities relating to climate change, 19 November 2007, available at: <http://daccessdds.un.org/doc/UNDOC/GEN/N07/464/25/PDF/N0746425.pdf?OpenElement>.
[59] For a good overview of the state of play of these discussions, as well as the various options under discussion, See UN CEB high-Level Committee on Programmes, 'UN system coordination arrangements on climate change, Issues Paper', 29 February 2008, CEB/2008/HLCP-XV/CRP.7, available at <http://hlcp.unsystemceb.org/documents/HLCP15th/CRP-7-ClimateChange.pdf>.
[60] See also the press release from the Government of Indonesia, 'informal trade minister dialogue on climate change', 10 December 2007, available at <http://www.menlh.go.id/popup.php?cat=17&id=2771&lang=en>.

they mostly provided a first exploratory debate on these two issues, among participants that were until recently fairly unfamiliar with the subject. In Bali, the upcoming Polish presidency of the UN climate change negotiations set to take place in Poznán, Poland, in December 2008, however already announced that they wish to continue both the finance and trade ministers' meetings in the context of the Poznán meeting.

Certainly the issue of financing and to some extent also the issue of trade are key in finding a solution to the climate change challenge. Following the first discussions at the Bali meeting and with a longer lead-time to prepare these meetings, there is a good opportunity for deepening the discussions in Poland and strengthening the ability of these meetings to come up with concrete results.

Discussions on the inter-linkage between trade and climate change may however be challenging because of the danger that the lack of progress in the Doha trade negotiations may spill over into these talks and the temptation to use these meetings to address contentious environment related trade measures introduced by developed countries, including for instance on biofuels, in particular palm nut oil and its link to deforestation. An additional challenge will be the added value that this new forum would bring compared to ongoing bilateral and multilateral trade discussions, including those in the context of the WTO. If the agenda could however continue to focus on issues on which a more constructive discussion will be possible, for instance looking at barriers to the trade in environmental goods as well as positive trade incentives to benefit climate change, some interesting progress could be made.

Future discussions between finance ministers in a follow-up to the Bali meeting could provide real added value to the two main finance discussions that are already ongoing. The first of these concerns the discussions within the context of the Gleneagles Dialogue and the work undertaken by the World Bank and other IFIs to implement the Gleneagles Plan of Action, which has so far mainly involved environment and energy ministries. The second discussion on financing was kicked off by a number of requests from Parties to the UNFCCC Secretariat to review and analyse investment and financial flows relevant to the development of an effective and appropriate international response to climate change, with particular focus on developing countries' needs, including their medium- to long-term requirements for investment and finance, at the UN climate meeting in Nairobi in November 2006.[61] The UNFCCC Secretariat decided to prepare a single report in response to these requests. A first draft of this report, prepared on the basis of a number of studies and input from experts

---

[61] See paragraph 8 of Decision 2/CP.12, FCCC/CP/2006/5/Add.1, available at: <http://unfccc.int/resource/docs/2006/cop12/eng/05a01.pdf>; and paragraphs 61 and 62 of the Report of the co-facilitators of the dialogue on long-term cooperative action to address climate change by enhancing implementation of the Convention, FCCC/CP/2006/5, page 15, available at <http://unfccc.int/resource/docs/2006/cop12/eng/05.pdf>.

and parties at three workshops, was presented at the inter-sessional UN climate change meeting in Vienna in August 2007 and the final report was presented in Bali.[62] The report was discussed in various contexts, and the Decision on the Bali Action Plan (discussed in more detail below) that launches negotiations on a post-2012 agreement explicitly refers to the 'enhanced action on the provision of financial resources and investment to support action on mitigation and adaptation and technology cooperation' as one of the issues to be addressed during these negotiations.[63] In developing the work programme under this Decision Parties will need to decide how to link in particular the discussions on financing with ongoing processes outside the UNFCCC, including the Gleneagles Dialogue and its follow-up, as well as a possible successor to the finance ministers' meeting.

A key question that will need to be resolved before the possible follow-up meetings in Poznán is what added value these two meetings can bring in addition to ongoing processes and how the meetings can be structured to provide that added value. The fact that, for the first time, finance and environment ministers themselves were engaged in these discussions is however encouraging and merits following up.

E. Mainstreaming: Challenges Ahead

Over the last few years the number of international processes and institutions that have started addressing climate change has increased enormously. This reflects not only the major political attention on this issue, but also a broader realization of the fact that climate change, both its causes and its impacts, is a broader sustainable development issue. Tackling climate change cannot be done in an isolated manner, but only through a full integration of climate change issues across the broad spectrum of sustainable development issues, including energy, transport, trade, economic and investment policies, and health.

Doing so however also brings a number of new challenges. Firstly, there is the need for a 'main platform' that can guide and facilitate the different strands of international cooperation work and keep pushing the agenda forward. The discussions under the UNFCCC and its Kyoto Protocol can and should continue to fulfil this role, in particular in the absence of an agreement of further international action 'post-2012'.

A second challenge is that a further diversification can be used to re-open or pre-empt decisions that were previously taken in the context of the UNFCCC

---

[62] The UNFCCC Secretariat's final report, 'Investment and financial flows to address climate change' as well as an impressive array of background documentation is available at <http://unfccc.int/cooperation_and_support/financial_mechanism/items/4053.php>.
[63] Paragraph 1(e) of Decision 1/CP.13 on the Bali Action Plan, and advance unedited version of which is available at <http://unfccc.int/files/meetings/cop_13/application/pdf/cp_bali_action.pdf>.

and its Kyoto Protocol, undermining the ability of those platforms to further elaborate common guidance and vision on global action to tackle climate change. Such erosion of the role of those platforms would be to the detriment of the effectiveness of international cooperation in tackling climate change and its impacts. The G-8 and the MEM are two good examples of platforms where this danger is most tangible. In view of the breadth of the climate challenge it is important that the key elements of international action to tackle this challenge have the broadest possible support among countries and address the wide range of concerns that are out there. This means not only involving those that are responsible for the largest share of global emissions, but also those that will suffer most from the impacts of climate change and those responsible for providing the fossil fuels that are ultimately responsible for those emissions. Obtaining support for global action across this broad set of interests can be slow and frustrating, and the temptation to 'sort this thing out among those that really matter' may be large. A limited engagement of countries will however only lead to solutions that are likely to be harder to implement in the long term across the broad range of actors involved. But that does not mean that smaller groups and platforms do not have a role: they can, and should, support the broader agenda and the discussions under the UNFCCC and its Kyoto Protocol.

A third challenge is one of coordination and awareness. Climate change remains relatively new and unknown for many of the new organizations that are getting engaged in this issue, with a great danger of mistakes or efforts to reinvent the wheel. Only solid coordination and information exchange between all institutions involved can avoid problems and ensure that the various work streams support rather than undermine each other.

5. The Bali UN Climate Change Conference

A. A Heavy Agenda

With the long list of high-level events preparing for the Bali meeting, the expectations for the outcome of this meeting were high. The Bali UN climate change conference, held from 3 to 15 December in Bali, Indonesia, combined a number of international climate negotiating processes. The most anxiously watched of these negotiations was the 13th Conference of the Parties of the UNFCCC (COP-13). The key issue on a full COP-13 agenda was the follow-up to the 'Convention Dialogue'. The Convention Dialogue was an informal dialogue among Parties to the UNFCCC that was started at COP-11 in Montreal under which Parties committed themselves to 'to engage in a dialogue, without prejudice to any future negotiations, commitments, process, framework or mandate under the Convention, to exchange experiences and analyse strategic approaches for long-term cooperative action to address climate change', however under the condition that this Dialogue 'will not open any negotiations leading to new commitments'.[64] The last of the four sessions of this Dialogue had taken place at the inter-sessional meeting in Vienna, Austria, in August 2007 and COP-13 was now presented with the report of the Dialogue's co-facilitators[65] and mandated to consider its results.

In addition to COP-13, there were however a number of other meetings with a substantial post-2012 agenda. Bali also hosted the 3rd Conference of the Parties serving as the meeting of the Parties to the Kyoto Protocol (CMP-3). CMP-3 had a range of issues on the implementation of the Kyoto Protocol on its agenda, including the finalization of the modalities for the Adaptation Fund, a fund set up to assist developing countries in their efforts to adapt to the increasing impacts of climate change.[66] The key post-2012 issue on the agenda of CMP-3 was the preparation of the 2nd review of the Kyoto Protocol, which is to take place at CMP-4 in Poznán in December 2008. The third strand of post-2012 oriented work was the resumed 4th session of the ad hoc working group on further commitments for Annex I Parties under the Kyoto Protocol (AWG-4). The work of this group, also set up at the UN climate change meeting in Montreal in 2005, was an important complement to that of the Convention Dialogue. A key difference is that, unlike the Convention Dialogue, it is actually engaged in concrete negotiations and is looking explicitly at further commitments for developed countries under the Kyoto Protocol. The US is however not participating in the work of this group, not being a Party to the Kyoto Protocol. AWG-4 had started its work at the inter-sessional meeting in Vienna in August, where it had discussed mitigation potentials of developed countries and the overall level of ambition for emission reductions by those countries by 2020. The main task on the agenda of the resumed AWG-4 in Bali was to agree a timetable for its further work. These three main strands of work (COP-13, CMP-3, AWG-4) were supported by sessions of the Subsidiary Bodies to the UNFCCC and its Kyoto Protocol as well as a wide range of side-events, and flanked by the finance and trade ministers' meetings discussed in the previous section.

The result of this heavy agenda was a dense two-week series of intense negotiations, in particular in the second week, with many meetings that continued late into the night or even into the early hours. The final plenary session only ended in the late afternoon of Saturday 15 December after more than 50 hours of almost uninterrupted negotiations. With the unprecedented public attention, the highest number of participants ever to a UN climate change meeting (more than

---

[64] Dialogue on Long-Term Cooperative Action to Address Climate Change by Enhancing the Implementation of the Convention, Decision 1/CP.11, FCCC/CP/2005/Add.1, p. 3, available at <http://unfccc.int/resource/docs/2005/cop11/eng/05a01.pdf>.
[65] Report on the dialogue on long-term cooperative action to address climate change by enhancing implementation of the Convention, Note by the co-facilitators, FCCC/CP/2007/4 of 19 October 2007, available at: <http://unfccc.int/resource/docs/2007/cop13/eng/04.pdf>.
[66] See the advance unedited version of Draft Decision -/CMP.3, Adaptation Fund, available at <http://unfccc.int/files/meetings/cop_13/application/pdf/cmp_af.pdf>.

10,000), and the high-level attendance during the closing session,[67] the expectations for the meeting were high. It is important to underline that the meeting was never supposed to agree on a post-2012 climate change agreement. The vast majority of Parties participating in the meeting did however expect it to deliver an agreement to commence discussions on a post-2012 agreement, with an ambitious timetable, a process for getting there and identifying the key issues to be explored to get to such agreement. The focus of this meeting was therefore on the negotiations on the follow-up to the Convention Dialogue. This was particularly so since the US was participating in those negotiations and since negotiations on further targets for developed Kyoto Parties are already underway and a 2nd review of the Kyoto Protocol is already planned for the end of 2008.

In the months leading up to the Bali conference, the EU had played a key role in raising the level of ambition and setting the agenda for the post-2012 discussions. The EU's position was underpinned by its climate change and energy package, agreed in March 2007 and discussed above. At the October EU Environment Council, the EU had also listed its key objectives for the Bali meeting,[68] central to which was the objective to reach agreement to launch negotiations of a global and comprehensive agreement to underpin effective global action on climate change after 2012, when the first commitment period of the Kyoto Protocol ends. The EU also wanted to set a clear deadline for the completion of those negotiations: the end of 2009, at the UN climate change conference in Copenhagen. As guidance for the negotiations the EU furthermore argued for the need to agree on a shared vision to tackle climate change that would limit the global average temperature increase to 2°C above pre-industrial levels (requiring global GHG emissions to be reduced to at least 50% below 1990 levels in 2050). The EU had also proposed a number of building blocks for a post-2012 agreement.[69]

B. A Difficult Final Spurt

The final sessions of COP 13 and CMP 3 were convened with a one-day delay on Saturday 15 December at eight in the morning and only finished at six in the

---

[67] Former Vice-President and Nobel Laureate Al Gore gave a speech on the Thursday 13 December and both UN Secretary General Ban Ki-moon as well as the Indonesian President attended and spoke at the final session on Saturday 15 December.

[68] Conclusions of the Environment Council of 30 October 2007, Climate Change: Preparations for the 13th session of the Conference of the Parties (COP 13) to the United Nations Framework Convention on Climate Change (UNFCCC) and the 3rd session of the Meeting of the Parties to the Kyoto Protocol (CMP 3) (Bali, 3–14 December 2007), available at <http://www.eu2007.pt/NR/rdonlyres/1CAB3779-074F-44D8-937F-C76ADEECD612/0/96899.pdf>.

[69] The EU's proposed building blocks were: deeper absolute emission reduction commitments by developed countries; facilitating further fair and effective contributions by other countries; extending the carbon market; increasing research, cooperation, development, and deployment of clean technologies; enhancing efforts to address adaptation; addressing emissions from international aviation and maritime transportation; reducing emissions from deforestation. In addition to this, the EU underlined the need for discussions on redirecting and scaling up investment in and finance for low-carbon technology.

evening. The meeting was loaded with emotions, drama, and surprise.[70] The COP President, the Indonesian Environment Minister Rachmat Witoelar, tabled a draft COP decision on the follow-up of the Convention Dialogue as a compromise, negotiations having been stopped after two days and nights without reaching a full agreement in the early hours of Saturday. The Group of 77 and China, representing the developing countries, could not accept the text and proposed an amendment that reformulated the paragraph on developing countries' action.[71] Following the proposal, the meeting was suspended at the insistence of the Chinese delegation, because the Indonesian Minister of Foreign Affairs was holding parallel informal consultations with Ministers from G77, of which neither the COP President nor the UNFCCC Secretariat were informed. An attempt to resume the session about an hour later failed, with an upset Chinese delegate for a second time pointing out that these consultations were still ongoing and openly accusing the Secretariat of convening two meetings in parallel in order to push a decision. When the plenary finally restarted, the President of Indonesia and the UN Secretary General joined the meeting and in passionate speeches appealed to all Parties to come to an agreement. The EU was the first to break the ice by accepting the textual amendment proposed by the developing countries. The US however refused to accept the amended text, because the references to developing countries' commitments were too weak for them. After a protracted and often emotional discussion, including forceful interventions from South Africa and Brazil, both underlining that 'developing countries are saying that we are willing to commit ourselves to measurable, reportable and verifiable mitigation actions',[72] the US dropped its opposition and accepted the text, opening the way for the agreement on the Bali Action Plan.

C. The Bali Action Plan

The EU achieved its main objective for Bali: COP-13 reached agreement on launching intensive negotiations on a global and comprehensive agreement for effective global action on climate change after 2012, when the first commitment period of the Kyoto Protocol ends, and ending these negotiations in 2009. The newly established Ad Hoc Working Group on Long-term Cooperative Action under the Convention (AWGLCA) will now operate alongside the Ad-Hoc Working Group (AWG) under the Kyoto Protocol that had already been created at CMP 1 in Montreal, but for which Parties in Bali also agreed to a 2009 end

---

[70] See for a more in-depth description of the events on the last night and day in Bali: Müller, B., 'Bali 2007: On the road again!' Oxford Energy and Environment Comment, February 2008, available at: <http://www.oxfordenergy.org/pdfs/comment_0208-2.pdf>.

[71] See infra for a more elaborate discussion of the different positions on the paragraph on developing country action.

[72] Oral intervention of the South African Minister Van Schalkwyk during the final plenary in Bali, see also Benito Müller, n.70 supra, p. 5.

date. Together, these two working groups provide a robust platform for negotiations on a post-2012 agreement.

There was however one key point on which the EU did not meet its objective for Bali: the level of ambition. Resistance of in particular the US, but also Russia and Japan prevented an agreement on such a level of ambition. Apart from the fact that the US did not have a position on what level of ambition should be pursued, its opposition can also be explained by the fact that it had explicitly set up the MEM to agree on this level by the end of 2008, with the Japanese having a similar expectation for their G-8 Presidency.

Being the result of a difficult negotiating process, the various texts resulting from the Bali meeting are riddled with ambiguities and sometimes leave space for interpretations that would imply only minimal further action and could even undermine key achievements of the multilateral agreements to tackle climate change as they stand.[73] In the light of the tremendous forward political momentum on climate change and the fact that the Bali outcome was the result of a consensus process that, in view of the strong positions of a small number of Parties, reflected the 'lowest common denominator', it can however be argued that the Bali outcome was a considerable breakthrough. This is not least because the meeting has actually kicked off negotiations on a post-2012 agreement and it is the first time since the start of the Bush administration that the United States have committed themselves to engage in those negotiations, and this on the basis of a decision that identifies all elements to be explored and creates a process to enable those negotiations to be successful. It is also because it is the first time that developing countries have agreed to discuss measurable, reportable, and verifiable actions to mitigate their GHG emissions.[74]

Three elements of the Bali Action Plan are worth exploring further in the context of this contribution: the shared vision, further action by developed countries, and further action by developing countries.[75]

D. A Shared Vision

Paragraph 1(a) of the Bali Action Plan requires Parties to address 'a shared vision for long-term cooperation, including a long-term global goal for emission reductions, to achieve the ultimate objective of the Convention, in accordance with the provisions and principles of the Convention, in particular the principle of common but differentiated responsibilities and respective capabilities, and taking into account social and economic conditions and other relevant factors'.

---

[73] See in particular also Rajamani, L., 'Berlin to Bali and Beyond: Killing Kyoto Softly?', forthcoming.
[74] See also the intervention of the South African Minister Van Schalkwyk, n. 72 supra, who even said that developing countries saying voluntarily that they are willing to commit to measurable, reportable, and verifiable mitigation actions 'has never happened before. A year ago, it was totally unthinkable.'
[75] For a much more detailed legal analysis of the Bali Action Plan see Rajamani, L., 'Berlin to Bali and Beyond: Killing Kyoto Softly?' n. 73 supra.

The concept of a 'shared vision' is not new. Apart from discussions on this in the context of the G8 and the MEM, it has been frequently used in the context of the AWG. The first part of AWG 4 in Vienna in 2007 even 'recognized that the contribution of Working Group III to the AR4 indicates that global emissions of greenhouse gases need to peak in the next 10 to 15 years and be reduced to very low levels, well below half of levels in 2000 by the middle of the twenty-first century in order to stabilize their concentrations in the atmosphere at the lowest levels assessed by the IPCC to date in its scenarios'.[76] The reference in the Bali Action Plan is however the first time that it is referred to in a UNFCCC text, i.e. with the support of the US.[77]

This reference to the shared vision also comes in combination with the last preambular paragraph which recognizes that 'deep cuts in global emissions will be required to achieve the ultimate objective of the Convention and emphasizing the urgency to address climate change as indicated in the Fourth Assessment Report of the Intergovernmental Panel on Climate Change'. Most interesting is the obscure footnote that is placed with the word 'urgency'. This footnote, referring to the specific elements of the Working Group III contribution to the AR4, was the result of protracted negotiations on the EU's proposal to include an explicit reference to the level of ambition. The footnote does not set such a level of ambition, but it does link a message on deep cuts with information on the different stabilization scenarios in the IPCC's AR4 and, most importantly, the following passage in the IPCC technical summary:

Under most equity interpretations, developed countries as a group would need to reduce their emissions significantly by 2020 (10–40% below 1990 levels) and to still lower levels by 2050 (40–95% below 1990 levels) for low to medium stabilization levels (450–550ppm CO_2-eq). Under most of the regime designs considered for such stabilization levels, developing-country emissions need to deviate below their projected baseline emissions within the next few decades.[78]

When taking this passage together with the messages from the G-8 meeting in Heiligendamm,[79] as well as the conclusions of the first part of AWG in Vienna,

---

[76] See the Report of the Ad Hoc Working Group on Further Commitments for Annex I Parties under the Kyoto Protocol on the first part of its fourth session, held at Vienna from 27 to 31 August 2007FCCC/KP/AWG/2007/4, para. 18, available at <http://unfccc.int/resource/docs/2007/awg4/eng/04.pdf>.

[77] The reference to a shared vision has however been frequently used in the conclusions of the AWG and has appeared in virtually all AWG conclusions since AWG 2 (6–14 November 2006). AWG 2 agreed that 'its work on further commitments by Annex I Parties should be guided by a shared vision of the challenge set by the ultimate objective of the Convention, based on the principles and other relevant provisions of the Convention and its Kyoto Protocol', para. 16 of FCCC/KP/AWG/2006/4, available at: <http://unfccc.int/resource/docs/2006/awg2/eng/04.pdf>.

[78] IPCC AR4 Working Group 3, technical summary, page 90, available at <http://www.ipcc.ch/pdf/assessment-report/ar4/wg3/ar4-wg3-ts.pdf>.

[79] See n. 46, as well as the discussion in the section 'The G-8: from Gleneagles via Heiligendamm to Hokkaido' supra.

one could conclude that the first elements of an agreement on a 'shared vision' are emerging, which are edging towards the lowest stabilization scenarios in the IPCC's AR4.[80]

E. Further Commitments for Developed Countries

Paragraph 1(b)(i) of the Bali Action Plan requires Parties to address in their negotiating process 'enhanced national/international action on mitigation of climate change, including, inter alia, consideration of':

> Measurable, reportable and verifiable nationally appropriate mitigation commitments or actions, including quantified emission limitation and reduction objectives, by all developed country Parties, while ensuring the comparability of efforts among them, taking into account differences in their national circumstances.

Of the various paragraphs in the Bali Action Plan, this paragraph, in combination with the developing country paragraph discussed below, was probably the most contentious in the negotiations. It was also significantly watered down from earlier proposals as the Bali negotiations edged forward.

The EU had initially wanted an explicit reference in the text to the need for developed countries 'as a group to reduce emissions in the range of 25–40% below 1990 levels by 2020'. This text was in fact heavily based on text that had already been agreed at the first part of AWG 4 in Vienna in August 2007.[81] Surprisingly, resistance to this paragraph did not only come from the US, but also from Russia, Japan, and Canada, as well as, surprisingly, a number of developing countries. The latter feared the context of this reference, which is the lowest IPCC emission scenario from the AR4 that explicitly also requires developing countries to reduce their emissions compared to 'business as usual'.[82]

In addition to this, the initial proposal to explicitly refer to mitigation commitments from developed countries was replaced by 'commitments or actions', which would 'include quantified emission limitation and reduction objectives'. These references are arguably the least ambitious in the Bali Action Plan and could well turn out to be an Achilles heel, not only of the Bali Action Plan, but also key premises of the UNFCCC and its Kyoto Protocol. By referring to 'commitments or actions' rather than 'commitments', that could 'include' rather than require 'quantified emission limitation and reduction objectives', the text opens the door to reconsidering the Kyoto Protocol's absolute emission

---

[80] See however also: Rajamani, n. 73 supra, who states that 'at this time, however, there is little of the vision that is shared'.

[81] See para. 19 of the AWG 4 conclusions, n. 75 supra, which reads: 'The AWG recognized that the contribution of Working Group III to the AR4 indicates that achieving the lowest stabilization level assessed by the IPCC to date and its corresponding potential damage limitation would require Annex I Parties as a group to reduce emissions in a range of 25–40 per cent below 1990 levels by 2020, through means that may be available to Annex I Parties to reach the emission reduction targets.'

[82] See also Rajamani, n. 73 supra.

reduction target based approach and replacing it by a bottom-up collection of nationally defined actions, without clear commitments.[83] This text is in particular worrying in view of the fact that the AWG is already discussing further reduction commitments for developed countries and has even discussed ranges for those commitments.[84] The text in the Bali Action Plan would allow the discussions that will now be held in the AWGLCA to reopen the 'acquis' of the AWG under the Kyoto Protocol and choose a completely different approach, in particular since the reference in the Bali Action Plan refers to 'all developed countries' and is not limited to those that are not a Party to the Kyoto Protocol.

Important for some of the EU Member States and EU industry was however the inclusion in this paragraph of an explicit reference to the 'comparability of efforts' among developed countries. Although the EU has already taken an independent commitment to reduce its greenhouse gas emissions by at least 20% by 2020 compared to 1990 levels, there continue to be concerns that this should be matched with comparable efforts in other developed countries, in particular if the EU should decide to further strengthen this commitment to a 30% reduction in the context of a post-2012 agreement. The reference to 'comparability of efforts' could be seen as a clear statement that all developed countries are expected to make a considerable effort. It could however also be seen as an argument of other developed countries to reduce their willingness to make an effort if one of them is unwilling to commit.

Overall, while the text on further action by developed country Parties could have been much more ambitious and better worded, it should not be forgotten that the Bali Action Plan only sets out the parameters for the negotiations. It does not define the outcome of those negotiations, which will be agreed on the basis of consensus and involve all Parties to the Convention. The wide parameters agreed in Bali respond to a very small number of Parties, in particular the US, supported by others whose position was mostly inspired by that of the US. Being only at the beginning of those negotiations and with the high likelihood that the US position will change following the US presidential elections at the end of 2008, it is too early to predict how Parties will use this wide mandate for the discussions.

6. Actions by Developing Countries

Protracted discussions on the paragraph on developing country action not only kept the negotiations going for most of the last night in Bali, but were also the main reason that the plenary discussions lasted into the late afternoon of Saturday

---

[83] Lavanya Rajamani, ibid, underlines this danger, inspiring the title of her article: 'Killing Kyoto softly?'
[84] n. 80 supra.

15 December. When no agreement could be reached during the last night, the chairs of the small group in which this text was being discussed took it back to the Indonesian COP President, who decided to put forward a compromise text to the Saturday morning plenary. Surprisingly this compromise text contained language that had already been rejected by the developing countries during the negotiations in the small group, which made it very unlikely that the group would change its views during plenary.[85]

Under the *chapeau* text that 'decides to launch a comprehensive process to enable the full, effective and sustained implementation of the Convention through long-term cooperative action, now, up to and beyond 2012, in order to reach an agreed outcome and adopt a decision at its fifteenth session, by addressing, inter alia [. . .] enhanced national/international action on mitigation of climate change, including, inter alia, consideration of [. . .]', the President had proposed the following paragraph:

measurable, reportable and verifiable nationally appropriate mitigation actions by developing country Parties in the context of sustainable development, supported by technology and enabled by financing and capacity building.

Perhaps most surprising was how close this text was to that proposed for the developed countries. There were only two significant differences. The first one was that for developing countries only 'actions' would be considered, whereas for developed countries this would include 'commitments or actions, including quantified emission limitation and reduction objectives'. The second one was that the developing country actions would need to be 'supported and enabled by technology, financing and capacity-building, in a measurable, reportable and verifiable manner', however without referring to who would provide this support. Much of this parallelism had been the result of pressure by the US, who were insisting on developing country action as a precondition for their agreement to engage in the negotiations and were negotiating hard to ensure similarity between the texts.

During the plenary session on Saturday, G-77 and China however proposed to amend the Chair's text to read as follows:

nationally appropriate mitigation actions by developing country Parties in the context of sustainable development, supported and enabled by technology, financing and capacity building, in a measurable, reportable and verifiable manner...

Although the EU gave its support for this new proposal, the US initially opposed this change. The main reason was that the US felt that the changed text watered down the actions by developing countries, being less explicit on their need to be measurable, reportable, and verifiable. Importantly, the US was also concerned by the possibility to interpret the text so that the 'measurable, reportable and

---

[85] For a detailed description of the discussions and statements of Parties See Mueller, n. 70 supra.

verifiable manner' would only apply to the 'supported and enabled by technology', and not to the 'mitigation actions'. Although the text does not explicitly state that this support would need to come from developed countries, the US feared that it would be placed in a situation where only developed countries would be expected to provide this support, and then in a 'measurable, reportable and verifiable manner'. It was only following the interventions by in particular South Africa and Brazil, that the US eventually decided to accept the proposal from G-77 and China, having 'especially listened to what has been said in this hall today, and that we are very heartened by the comments and the expression of firm commitments that have in fact been expressed by developing countries'.[86]

One key point in the text of the Bali Action Plan that has received less attention so far is that it makes an important departure from the UNFCCC and its Kyoto Protocol in its use of the terminology 'developed' and 'developing' country Parties. The UNFCCC divided the world into three categories: Annex II consists of the members of the OECD in 1992; Annex I, consisting of that same list plus the economies in transition; and the rest of the group. When the Berlin Mandate was agreed in 1994, providing the mandate for the start of the negotiations on the Kyoto Protocol,[87] it explicitly referred to the 'strengthening of commitments of the Parties included in Annex I to the Convention'. Rather than referring to Annex I, the Bali Action Plan however refers to 'developed countries', which could be interpreted to suggest that the previous split between Annex I and non-Annex I countries may be open for reinterpretation in the future.

Most developed countries, but also a number of developing countries, would welcome opening up the 'rigid' three-category structure of the UNFCCC and introduce a wider differentiation among countries, in particular also among developing countries. The US has made this point on various occasions during the past few years and used this as one of the reasons for starting the MEM. Also the EU would welcome such differentiation, as is apparent from its explicit statement that it is willing to increase its commitment to reduce its greenhouse gas emission to 30% provided not only that 'other developed countries commit themselves to comparable emission reductions' but also that 'economically more advanced developing countries to contributing adequately according to their responsibilities and respective capabilities'.[88]

Interestingly, during the final session in Bali, Bangladesh sought to include a reference to 'differences in national circumstances' into the developing country paragraph, seeking recognition for the plight of the particularly vulnerable and

---

[86] Ibid. p. 5.
[87] Decision 1/CP.1, The Berlin Mandate: Review of the adequacy of Article 4, paragraph 2(a) and (b), of the Convention, including proposals related to a protocol and decisions on follow-up, FCCC/CP/1995/7/Add.1, p.4, 7 April 1995, available at <http://unfccc.int/resource/docs/cop1/07a01.pdf>.
[88] See inter alia paragraph 11 of the Conclusions of the 30 October EU Environment Council, n. 68 supra.

poor developing countries that also have very low emissions and lack the rapid economic growth of their emerging brethren. This insertion was however rejected by India and China and did not make it into the final text. Differentiation within the group of developing countries has thus far been fiercely opposed by G-77 and China.[89] In view of the growing political pressure both from outside and within the group, but also the changing economic realities in today's world, the internal differentiation within the group of developing countries is however very likely to be one of the key issues for the upcoming negotiations.

Such discussions will be extremely complicated, not in the least because of the enormous differences in the type and level of development of countries such as India, China, and Brazil. In order for them to succeed it will be vital that these discussions not only address the mitigation expectations for more advanced developing countries, but in particular also address the set of instruments to help enable them to meet those expectations. In doing so it is vital that those instruments respect the particular characteristics of individual countries—China and India's level and type of economic development are for instance difficult to compare. The text of the Bali Action Plan provides a number of important openings for discussions on instruments to assist developing countries in their mitigation efforts. The developing country paragraph explicitly refers to mitigation actions by developing countries that are 'supported and enabled by technology, financing and capacity building'. In addition to that, the text also contains important sections on 'enhanced technology development and transfer', as well as 'enhanced action on the provision of financial resources and investment'. A meaningful outcome of the discussions on the latter two will be a *sine qua non* for a strengthened engagement of the more advanced developed countries.

7. Conclusion: The Road from Bali

Over the last few years we have observed a landslide change in the international climate negotiations, culminating in the outcome of the Bali UN climate conference. We are however still only at the beginning of what will not be an easy negotiating process that is set to operate on a very ambitious timetable, aiming to be finalized before the end of 2009.

Among the multiple challenges that negotiators will be faced with over the period ahead, there are four key challenges that are worth highlighting: forging a consensus on the shared vision for international action on climate change; strengthening the commitment of developed countries to further Kyoto-like emission reduction targets; defining the expectations for developing country participation; and deciding how the necessary finance and investment can be

---

[89] More elaborately, Rajamani, n. 73 supra.

mobilized to support international climate action, and in particular both adaptation and mitigation actions in developing countries.

The first two of these have thus far been central planks of the EU's post-2012 strategy. The EU has been actively working on forging a consensus on the global vision for international action on climate change, taking as a basis the EU's 2°C objective, but also translating this into concrete emission reduction objectives for 2050. As already discussed above, discussions on the shared vision are already taking place in the context of the AWG and will now also happen in the context of the AWGLCA. In addition to that, this issue will be picked up in the context of the MEM and the Japanese G-8 Presidency will place it high on the agenda for the G-8 summit as well. While agreement on such 'shared vision' will be key to guide the post-2012 negotiations and in particular the level of effort of Parties, its value should not be overestimated. Discussions should not focus on a firm long-term 'target', but get a common view on a vision or level of ambition that will provide the necessary guidance to the discussions on the efforts expected from different Parties under a future regime.

Strengthening the commitment of developed countries to further Kyoto-like emission reduction targets will be key. These targets, supported by the right type of instruments, including for instance the global carbon market, give the best guarantee for effective international action to reduce global greenhouse gas emissions. A continued use of such targets will also be important for the continuation and further growth of the global carbon market.

The EU has indicated its commitment to further Kyoto-like emission reduction targets, as have a number of other, mostly European, developed countries. Japan's Prime Minister Yasuo Fukuda at the World Economic Forum meeting in Davos in January 2008, not only announced that Japan's aim for the Hokkaido G-8 summit is 'the setting of fair and equitable emissions target', but also that 'within that context, Japan will, along with other major emitters, set a quantified national target for the greenhouse gas emissions reductions to be realized from now on'.[90] The failure of the G-8 to agree on a mid-term (2020) target and only propose a long-term target (2050) however undermined this effort. Although the G-8 did agree to 'implement ambitious economy-wide mid-term goals in order to achieve absolute emissions reductions', it remains unclear whether this means a firm commitment to determine and enforce such targets at the international level. More importantly, it remains unclear to what extent Russia and in particular the US are willing to (continue to) follow this approach. Much of this will depend on the next US administration, following the US presidential elections in November 2008. A clear signal from the next US administration that the US will not support quantified emission reduction objectives may well prove the end of the Kyoto

---

[90] Special Address by H. E. Mr. Yasuo Fukuda, Prime Minister of Japan on the Occasion of the Annual Meeting of the World Economic Forum, Congress Center, Davos, Switzerland, 26 January 2008, available at: <http://www.mofa.go.jp/policy/economy/wef/2008/address-s.html>.

Protocol approach, confirming fears that the wide negotiating margin set in the Bali Action Plan could be an important first opening to reconsider the multilateral approach to tackle climate change.

While on these first two challenges different views between countries have become increasingly clear, and there is a good basis for discussion, if not agreement, on the second set of challenges the discussion remains much more open. These two issues also present the clearest gap in the EU's post-2012 strategy as it has been defined in the EU's various strategy papers over the last two years. Much more analytical work remains to be done over the next year and a half to identify and choose concrete options.

As for the exact actions or contributions that developing countries will be expected to make, it is becoming increasingly clear that this will not be a 'one size fits all' solution. Any viable post-2012 agreement will need to come up with a catalogue of tools for strengthening developing country engagement, particularly so in the area of mitigation. This catalogue will need to reflect the growing differentiation within the group of developing countries, hence the importance of the political signals on the need for such differentiation in Bali. The lack of clear ideas, in particular also on the side of developing countries, on what those options could be has so far very much hampered the international climate debate. The fact that the focus of the debate so far has been on the global vision and the targets for developed countries has made developing countries very nervous about the commitments that they would be expected to take. The lack of clarity on their contribution, not helped by the outright refusal of a number of developing countries to discuss such contribution, has in particular also hampered discussions on the shared global vision. Key developing countries fear that an agreement on a shared global vision and developed countries determining their own contribution would imply that the resulting efforts would need to be made 'by default' by developing countries. Without further clarity on developing country contributions, reaching agreement on a shared global vision will be extremely challenging.

The latter then automatically brings us to the fourth challenge: finance and investment. This challenge is the most difficult, but also the most important: how do we mobilize the necessary finance and investment to enable us to reach the necessary deep cuts in global greenhouse gas emissions to avoid dangerous levels of climate change and to adapt to the impacts of unavoidable climate change? As already stated above, without more clarity on this challenge it will be extremely hard to engage developing countries in a meaningful debate on their efforts. The finance and investment debate is therefore pivotal in the post-2012 negotiations. This also entails an important governance challenge. Decisions on mobilizing the necessary finance and investment are not taken by environment ministries, but depend on a much fuller mainstreaming of climate change than we have seen to date and require the engagement of a different and broader type and level of discussion partners. The mainstreaming described above that we have seen so far is a good start. We have also seen a first set of attempts to get a finance and investment discussion going, under the Gleneagles Dialogue, among international financial institutions, and with the UNFCCC paper and finance ministers meeting on climate change during 2007. This debate must now urgently be taken forward, starting with an early and strengthened engagement of Heads of State and Government and finance and development ministers. In this debate, it will however be important to realize that this is not just about developed countries putting further official development assistance (ODA) money onto the table. ODA streams will continue to be vital and must be increased. But ODA alone cannot solve the climate challenge. Key will be the engagement of the private sector. ODA spending is dwarfed by private sector investment streams, including the budding global carbon market. The most recent World Bank Report on the developments in the still relatively new global carbon market for instance shows that the value of this market more than doubled in 2007 compared to 2006, rising to a value of over €47 billion, which is almost two-thirds of total global ODA.[91] The most difficult part will therefore be defining the necessary frameworks to steer private sector investments and enable the rapid further development and deployment of new technologies, in particular also in developing countries.

Underpinning all of these challenges is the need for success stories and companies and countries as well as individuals leading by example. Economic analysis continues to tell us that we can reduce global greenhouse gas emissions without wrecking our economies, and often even at considerable benefit. What remains missing is the political courage of countries to follow this economic advice and translate it into a robust domestic climate policy. The Commission's proposals from 23 January 2008 for the implementation of the EU's climate change and energy strategy provide an opportunity for the EU to take an important step in that direction. Those proposals must now be translated into concrete action.

---

[91] For an overview of the investment streams resulting from the global carbon market, see 'The World Bank, State and Trends of the Carbon Market 2008, Washington, D.C.–May 2008', available at <http://siteresources.worldbank.org/NEWS/Resources/State&Trendsformatted06May10pm.pdf>. Also see the World Bank's work on the new investment framework for clean energy and development, n. 41 supra. For a recent overview of global ODA streams, see the 'OECD Journal on Development: Development Co-operation Report 2007', available at <http://www.oecd.org/document/32/0,3343,en_2649_34485_40056608_1_1_1_37413,00.html>.

Index

access
  information, to, transparency and, REACH framework 89–90
accountability
  concerns about Common Implementation Strategy and 50–3, 54
  registration, in, REACH framework 63–6
acquis communautaire
  environmental policy and 1, 169
Action Plan *see also* Environmental Action Programme
  Bali UN Climate Change Conference 173, 197, 201–6
  Gleneagles 190–91
Adebowale, M. 93, 124
adjudication
  proceedings, substitution and 79
AETR doctrine
  ECJ, exclusivity question and 132–3
agenda *see also* international agenda
  Bali UN Climate Change Conference 198–200
Agreement on the Application of Sanitary and Phytosanitary Measures (SPS Agreement)
  trade and environment external relations and 153–6
Agreement on Technical Barriers to Trade (TBT Agreement)
  trade and environment external relations and 156–7
agreements *see also* Burden Sharing Agreement; General Agreement on Tariffs and Trade; General Agreement on Trade in Services; legislation; Multilateral Environment Agreements
  international environmental, EC external relations and 136
  mixed, EC external relations and 139–41, 145
ambition *see* objectives
artificial or heavily modified body of water
  good water status exception, Water Framework Directive 33–4
attribution of powers principle
  external relations and 128–30
authorization *see also* prior authorization
  contestability and, REACH framework
    criteria 76
    procedure 75–8
    substance requiring 69–71
  industry responsibility and, REACH framework 66–8
  provisionality and, REACH framework 80–82

Baden, B. M. 121
Bali UN Climate Change Conference *see also* climate change; United Nations
  Action Plan 201–6
  actions by developing countries 206–8
  agenda 198–200
  final sessions 200–1
  Finance and Trade Ministers' meetings 195–7
  future challenges 208–11
  IPCC Fourth Assessment Report adopted at 174
Barroso, José Manuel 184–5
Brazil
  international climate change policy progress and 176, 201
burden of proof
  authorization applications and, REACH framework 67
Burden Sharing Agreement (EU) *see also* agreements
  external relations and 128
  international climate change regime and 164, 167–8
business associations
  interests, factors affecting governance evolution 5, 7

Canada
  international climate change policy progress and 204
Capacity Global
  environmental justice and 93
Cardiff Process of environmental policy integration *see also* environmental policy
  sustainable development regime and 19–20, 23

Cartagena Protocol on Biosafety
  see also legislation
  ECJ Opinion 128, 138, 144–5
challenges
  future
    Bali UN Climate Change Conference and 208–11
    international policy agenda, mainstreaming climate change into 197–8
    key, identified by EU SDS 24
change *see also* climate change
  governance, limits on 24–6
  polity, governance evolution and 5–7
characteristics analysis
  River Basin Management Plan and 38
chemical safety report (CSR)
  access to information and, REACH framework 90
  authorization applications and, REACH framework 66
  registration requirements and, REACH framework 63
chemical status
  good water status and, Water Framework Directive 30, 32–3
Chemicals Agency *see also* European Union
  access to information and 88–90
  authorization and 70, 71, 75
  classification and labelling and 72
  contestability limitations and 74
  establishment, 56
  registration and 60, 63–5
  reporting/review/revision and, REACH framework 83
  restrictions and 76
  substance evaluation and 69
  substitution and 78–80
chemicals sector *see also* REACH framework
  Lisbon goals and 57
China
  international climate change policy progress and 176, 188, 191, 193, 201, 206–8
CIS *see* Common Implementation Strategy
classification
  substance
    harmonized, contestability and, REACH framework 71–2
    transparency and, REACH framework 86–8

climate change *see also* Bali UN Climate Change Conference; Kyoto Protocol on Climate Change; United Nations Framework Convention on Climate Change
  ESDP and 114–15
  external relations and 128
  governance change limits and 2
  international regime 161–5
  Kyoto Protocol non-compliance, international and EU responsibility issues 165–8
climate change policy *see also* EC Climate Change Policy; International Panel on Climate Change
  international agenda, mainstreaming climate change into 189–90
  challenges ahead 197–8
  Finance and Trade Ministers' meetings, Bali 195–7
  G-8 and 190–92
  major economies meeting 192–4
  UN and 194–5
  progress, EU and international 171–3
    energy security and competitiveness and, 183–9
    IPPC and politics 173–9
Coalition for Environmental Justice
  environmental justice and 93
cohesion *see also* integration
  environmental justice and, 119, 125
  ESDP and 115–18
Cohesion Fund
  ESDP and 96
command *see* control
Commission *see also* European Union
  authorization and restrictions procedures comparison and, REACH framework 77
  authorization procedure and, REACH framework 76
  Chemicals Agency as agent of 74
  CIS and 49, 50
  climate change policy progress and 172–3, 185–6
  cohesion policy and 119
  contestability and REACH framework and 72, 74
  environment regime and 9
  ESDP governance and 104–5
  ESDP origins and 99–100
  EU participation in work of IPCC and 180, 181

Index

ideological orientation, factors affecting governance evolution 5, 7
internal market regime and 13
opt-outs under EC Treaty Article 95 and, REACH framework 73
provisionality and, REACH framework 81
reporting/review/revision and, REACH framework 82–3
restrictions procedure and, REACH framework 77
River Basin Management Plan sent to 37
spatial planning and 96
substitution and 79–80
sustainable development regime and 19
Water Framework Directive and 42–3, 45–6, 55

common commercial policy
EC Treaty and 128, 142–4

Common Implementation Strategy (CIS) *see also* **strategy; Water Framework Directive**
activities 47–50
concerns about 50–55
explained 28, 46
information forcing and 36
organization 46–7
sustainable development regime 20–23

competence
concurrent, 135, 139
external, 127–30, 135, 143
implied external relations, ECJ doctrine of 130

competition *see also* **economics**
factors affecting governance evolution 4
integration regime and 15
internal market regime and 11–14

competitiveness
climate change policy and, EU and international progress in 183–9
ESDP and 103

compliance *see also* **non-compliance**
checking, industry accountability in registration and, REACH framework 63–4

consumer
substance classification and labelling and, REACH framework 86–7
transparency and, REACH framework 86

contestability
REACH framework core element 58, 68–9
authorization criteria 76
authorization procedure 75–8

classification and labelling, harmonized 71–2
EC Treaty (Article 95) 72–3
limitations 73–5
Member State safeguards 72
restrictions 71
restriction criteria 77
restriction procedure 76–8
substance evaluation 69
substance requiring authorization 69–71
'Territory of the EU' concept, of 111

control *see also* **obligations**
hierarchical, industry accountability in registration and, REACH framework 63–4
River Basin Management Plan and 38–9
Water Framework Directive and 28, 40–41

Council of Europe *see also* **European Union**
spatial approach begun by 94, 108

criteria
authorization, contestability and 76
restriction, contestability and 77

CSR *see* **chemical safety report**

data *see also* **information; Safety Data Sheet**
sharing, industry accountability in registration and, REACH framework 65–6

decision-making *see also* **justification; Member States**
environment regime 8–10
ESDP and 95, 104, 114
governance change limits and 24–5
integration regime 14–15
internal market regime 12
key feature of governance regimes 7
sustainable development regime 20–21
Water Framework Directive and 27–8, 36–8, 41–2

developed countries
climate change commitments for, Bali Action Plan (para 1(b)(i)) 205–6
meetings, international climate change policy agenda and 192–4
shared vision, Bali Action Plan (para 1(a)) 202–4

developing countries
climate change actions by, Bali UN Climate Change Conference and 205–8
international climate change policy progress and 188–9, 201

216 Index

developing countries *(cont.)*
  shared vision, Bali Action Plan (para 1(a)) 202–4
development *see* European Spatial Development Perspective; infrastructure development; sustainable development
duties *see* obligations

EAP *see* Environmental Action Programme
EC Climate Change Policy *see also* climate change policy
  external relations and 128
EC Treaty *see also* European Union; treaties
  attribution of powers principle (Article 5) 128–9
  contestability and (Article 95) 58, 68, 72–3
  common commercial policy and 128, 142–4
  environmental sector and (Article 174) 142, 144, 168
  external relations competence established by 127–9
  interpretation of Article 5 129–30
  loyalty principle, Article 10 134, 167
  spatial planning and 99
  territorial application, Article 299 136
ECJ *see* European Court of Justice
ecological status
  environmental objectives definition and, Common Implementation Strategy 48–9
  good water status and, Water Framework Directive 30–32
economic analysis of water use
  River Basin Management Plan and 38–40
economics *see also* competition
  climate change and, ESDP approach and 115
  environmental justice absence and, ESDP 118, 125
  governance evolution and 4
  sustainable development and, ESDP 106–7
  sustainable development regime and 19
effectiveness (necessary attainment) principle
  existence question and, ECJ 130–2
energy security *see also* International Energy Agency
  climate change policy and, EU and international progress in 183–9
enforcement
  REACH framework 84
environment

ESDP and 114–15
external relations in 127
governance evolution and 4
threats, environment regime and 10
trade and *see* trade and environment
environment regime *see also* governance regimes
  environmental policy politicization in Member States and 9–10
  governance evolution and 8–11
Environmental Action Programme (EAP) *see also* Action Plan; European Union
  1st, governance evolution and 8–9
  5th, integration regime and 16
  6th, sustainable development regime and 19–20
environmental conditionality
  external relations and 128
environmental justice
  absence of, ESDP and 118–24
environmental objectives *see also* objectives
  Common Implementation Strategy and 48–50
  Water Framework Directive 30
environmental policy *see also* Cardiff Process of environmental policy integration
  acquis communautaire and 1, 169
  emergence 2
  institutional layering and 3
  Member State politicization, environment regime 9–10
environmental protection
  external relations and
    attribution of powers principle, EC Treaty Article 5 128–9
    exclusivity question 128–9, 132–4, 138–41
    existence question 128–32, 138–41
  REACH framework and 91
environmental sector
  EC Treaty Article 174 142, 144
  exclusivity and existence questions in 138–41
  external relations and 134–38
ESDP *see* European Spatial Development Perspective
ESPON *see* European Spatial Planning Observatory Network
EU *see* European Union
EU Sustainable Development Strategy (EU SDS) *see also* European Union; strategy
  key challenges identified by 24

sustainable development regime measures and 23–4
European Court of Justice (ECJ)
AETR doctrine, exclusivity question 132–3
authorization and restrictions procedures comparison and, REACH framework 77
effectiveness principle, existence question 130–2
exclusivity principle, exclusivity question 133–4
existence and exclusivity questions in environmental field 138
external relations Opinions 128, 143–5
functional jurisdiction 136–7
implied external powers principle, existence question 130–31
implied external relations competence doctrine, existence question 130
lack of parallelism between EC internal and external competence principle 129
loyalty principle, exclusivity question 133–4
mandatory requirements doctrine, trade and environment 158–9
trade and environment and 142
European Environment Agency
environmental justice and 93
European Environment Bureau
Common Implementation Strategy and 50
European Spatial Development Perspective (ESDP)
elements 105–6
  cohesion 115–18
  environment 114–15
  mobility 111–14
  sustainable development 106–7
  territory 107–11
environmental justice, absence of 93–4, 118–24
governance and 104–5
impetus, origins and legal base 98–101
key objectives 102–4
main aim 92–3
spatial strategy and 94–8
sustainable development and environmental justice 'movements', importance of 124–6
'Territory of the EU' and 92–3
USA comparison 93
European Spatial Planning Observatory Network (ESPON)

environmental justice absence and 118–19, 124
ESDP governance and 104
'Territory of the EU' and 109
European Union (EU) *see also* individually-named member states; Commission; Council of Europe; external relations; Parliament
behaviour in WTO 157–61
climate change and energy strategy 187–8, 200
climate change policy progress and 184, 209–11
enlargement
  governance evolution and 5
  integration regime and 14, 16
  sustainable development regime and 19–20
GATT Article XX(b) application and 149–50
international climate change regime and 162–5
Kyoto Protocol non-compliance responsibility issues and 165–8
participation in work of IPCC 180–81
SPS Agreement and 154–6
TBT Agreement and 156–7
evaluation
substance, contestability and, REACH framework 69
exceptions
good water status, Water Framework Directive 33–5, 37–8
exclusivity principle
exclusivity question and, ECJ 133–4
exclusivity question
external relations and
  background 128–9
  environmental field, in 138–41
  examination 132–4
exemptions *see also* **opt-outs**
REACH registration requirements 60–61
existence question
external relations and
  background 128–9
  environmental field, in 138–41
  examination 129–32
external relations *see also* **European Union**
climate change
  international regime 161–5
  Kyoto Protocol non-compliance, international and EU responsibility issues 165–8

external relations *(cont.)*
  competence established by EC Treaty 127
  environmental field 127, 134–8
    exclusivity and existence questions in 138–41
  environmental protection and
    attribution of powers principle, EC Treaty Article 5 128–9
    exclusivity question 128–9, 132–4, 138–41
    existence question 128–32, 138–41
  trade and environment and 127–8
    common commercial policy, EC Treaty Article 133 142–4
    environmental sector, EC Treaty Article 174 142, 144, 168
    GATT 146–53
    MEAs 144–6
    sector described 141–2
    SPS Agreement 153–6
    TBT Agreement 156–7
    WTO context 146–8, 157–61

Faludi, A. 99, 100–101, 109–10
flexibility
  Common Implementation Strategy and 48
  Water Framework Directive 28, 35–7, 41
fragmentation
  water protection policy and 29
France *see also* European Union; Member States
  GATT Article XX(b) application and 149–52
  'Territory of the EU' and 108
freedom of movement
  mobility and 112
functional jurisdiction
  external relations and 136–7
funding programmes
  ESDP and 95–6
future
  challenges
    Bali UN Climate Change Conference and 208–11
    international policy agenda, mainstreaming climate change into 197–8
  IPCC 181–3

G-8
  meetings, international climate change policy agenda and 190–92

GATS *see* General Agreement on Trade in Services
General Agreement on Tariffs and Trade (GATT) *see also* agreements
  trade and environment external relations and 143–4, 146–8
  Article XX(b) 147–52
  Article XX(g) 147, 152–3
General Agreement on Trade in Services (GATS) *see also* agreements
  trade and environment external relations and 143
Germany *see also* European Union; Member States
  environment regime and 9–10
  integration regime and 17
  internal market regime and 14
  international climate change policy progress and 191–2
GHS *see* Globally Harmonized System of Classification and Labelling of Chemicals
Giannakourou, G. 98, 117, 119
globalization
  factors affecting governance evolution 4
  ESDP and 92, 94, 102–3
  integration regime and 15
  sustainable development regime and 19
Globally Harmonized System of Classification and Labelling of Chemicals (GHS) *see also* harmonization
  transparency and, REACH framework 86–7
good water status
  River Basin Management Plan and 37–8
  Water Framework Directive and 30–33
    exceptions 33–5
Gore, Al 172
governance
  ESDP and 104–5
  evolution 1–4, 24–6
    environmental and economic conditions affecting 4
    environment regime 8–11
    integration regime 14–18
    internal market regime 11–14
    international political developments and commitments affecting 4
    polity changes affecting 5–7
    sustainable development regime 18–24
  land use and development 104–5
  REACH framework approach 57–8, 90
  water protection 27–9, 55

governance regimes *see also* environment regime; integration regime; internal market regime; sustainable development regime
  key features 6, 7
  overview 3
  policy development and 1
governance techniques
  Water Framework Directive and
    economic instruments 40–42
    information forcing 36–40
    public participation 43–5

harmonization *see also* Globally Harmonized System of Classification and Labelling of Chemicals
  contestability and, REACH framework 68
  exclusivity question and 134, 138–9
  external relations in trade and environment and 145
  integration regime and 15
  internal market regime and 11, 13
  REACH framework achievements 91
  substance classification and labelling, contestability and, REACH framework 71–2
Healey, P. 97, 108–9
health *see also* pollution
  factors affecting governance evolution 4
  protection, REACH framework and 91
  threats, environment regime and 10
hierarchical control *see also* control
  industry accountability in registration and, REACH framework 63–4
human impact analysis
  River Basin Management Plan and 38–9
ideological orientation
  factors affecting governance evolution 5, 7

IEA *see* International Energy Agency
impetus
  ESDP 98–101
implementation
  REACH framework, importance attached to 58–9, 90–91
implied external powers principle
  existence question and, ECJ 130–31
implied external relations competence doctrine
  ECJ, existence question and 130
incentives
  registration and, REACH framework 60, 63
  substitution and, REACH framework 80

India
  international climate change policy progress and 176, 179, 183, 188–9, 191, 193, 208
industry
  accountability in registration, REACH framework 63–6
  information burden on, REACH framework authorization 66–8
  registration 60–3
industry responsibility
  REACH framework core element 58
  authorization 66–8
  registration 60–66
information *see also* data; public information; Safety Data Sheet; substance information exchange forum
  access to, transparency and, REACH framework 88–90
  burden on industry, REACH framework authorization 66–8
  registration 60–3
  generation and dissemination, REACH framework priority 57
  hazardous substances, transparency and disclosure of, REACH framework 85, 88
  regulatory tool, as, REACH framework 59
  supply chain, transparency and, REACH framework 84–6
  Water Framework Directive and 27–8, 33–4
information forcing
  Water Framework Directive and 36–40
infrastructure development
  mobility and 112–13
institutional layering
  governance change limits and 24–5
  governance regimes evolution and 3
instruments *see also* legislation
  economic, Water Framework Directive and 40–42
  environment regime 10–11
  integration regime 16–18
  internal market regime 13–14
  sustainable development regime 23
  type of, key feature of governance regimes 7
integration *see also* cohesion; regional policy; 'Territory of the EU'
  ESDP and 92–4, 96–7
  transport and 112
  Water Framework Directive and 29–30, 38

integration regime *see also* **governance regimes**
  governance change limits and 25
  governance evolution and 14–18
intercalibration
  environmental objectives definition and, Common Implementation Strategy 48–50
interests
  governance evolution and 5, 7
internal market
  ESDP and 95
internal market regime *see also* **governance regimes**
  governance evolution and 11–14
international agenda *see also* **agenda**
  climate change policy 189–90
    challenges ahead 197–8
    Finance and Trade Ministers' meetings, Bali 195–7
    G-8 and 190–92
    major economies meeting 192–4
    UN (United Nations) and 194–5
International Energy Agency (IEA) *see also* **energy security**
  international climate change policy progress and 190–91
International Panel on Climate Change (IPCC) *see also* **climate change policy**
  EU participation in work of 180–81
  Fourth Assessment Report 173–9, 184, 203
  future 181–3
  international climate change policy progress and 171–2, 203–4
  provisionality and 81
international politics *see also* **politics**
  developments and commitments affecting governance evolution 4
interpretation
  EC Treaty Article 5 129–30
INTERREG
  ESDP and 95, 101

Japan
  international climate change policy progress and 192, 202, 204, 209
Jensen, O. 111, 113, 115
joint registration *see also* **registration**
  industry accountability and, REACH framework 64–5
jurisdiction
  functional, external relations and 136–7
justice *see* **environmental justice**

justification *see also* **decision-making**
  contestability in substance evaluation and, REACH framework 69
  governance change limits and 26
  industry accountability in registration and, REACH framework 64
  integration regime 15–16
  internal market regime 13
  key feature of governance regimes 7
  substitution and, REACH framework 79
  sustainable development regime 22
  Water Framework Directive and 38

Kyoto Protocol on Climate Change *see also* **climate change; legislation; United Nations Framework Convention on Climate Change**
  EC and 163–5
  EC external relations and 128
  EU participation in work of IPCC and 180, 181
  international climate change policy progress and 172, 197–9, 201–2
  international climate change regime and 162
  non-compliance, international and EU responsibility issues 165–8

labelling
  substance
    harmonized, contestability and, REACH framework 71–2
    transparency and, REACH framework 86–8
land *see* **European Spatial Development Perspective; spatial strategy; territory**
legal base
  ESDP 98–101
  external relations in trade and environment and 145
  internal market regime 12
  key feature of governance regimes 7
  REACH framework 68
legislation *see also* **agreements; Cartagena Protocol on Biosafety; instruments; Kyoto Protocol on Climate Change; treaties; Water Framework Directive**
  governance change limits and 25–6
  internal EC, external relations and legitimacy of 136–8
  number of items adopted 2
legitimacy
  integration regime and 14–15

internal EC legislation, external relations and 136–8
limitations *see also* restrictions
   contestability, REACH framework 73–5
locally undesirable land uses (LULUs)
   environmental justice in USA and 122
   ESDP and 93
loyalty principle
   exclusivity question and, 133–4, 139
   international climate change regime and 167
LULUs *see* locally undesirable land uses

major economies *see* developed countries
mandatory requirements doctrine
   ECJ, trade and environment and 158–9
market *see* internal market; product market
Marks, G. 96–7
MEAs *see* Multilateral Environment Agreements
meetings *see also* Bali UN Climate Change Conference
   international climate change policy agenda and
      Finance and Trade Ministers, Bali 195–7
      G-8 190–92
      major economies 192–4
Member States *see also* individually-named member states; contestability; decision-making; European Union
   cohesion and 116
   environmental policy politicization, environment regime 9–10
   ESDP governance and 104
   EU participation in work of IPCC and 180–81
   integration regime emergence 14
   interests
      environment regime 9
      factors affecting governance evolution 5, 7
   IPCC Fourth Assessment Report and 176
   process standards, internal market regime 12
   product standards, internal market regime 13
   reporting/review/revision and, REACH framework 83–4
   River Basin Management Plan and 38–40
   safeguards, contestability and, REACH framework 72
   'Territory of the EU' and 109
   Water Framework Directive implementation and 41–2
mobility *see also* trans-European transport network

environmental justice absence and 119–20
   ESDP and 111–14
mode of justification *see* justification
Multilateral Environment Agreements (MEAs) *see also* agreements
   trade and environment external relations and 144–6

necessary attainment principle *see* effectiveness principle
necessity principle
   GATT Article XX(b) application and 149–50, 152
   trade and environment case law and, ECJ and WTO 159–60
networks *see also* European Spatial Planning Observatory Network; trans-European transport network
   ESDP governance and 104
NGOs *see* non-governmental organizations
non-compliance *see also* compliance
   Kyoto Protocol, international and EU responsibility issues 165–8
non-governmental organizations (NGOs)
   interests, factors affecting governance evolution 5, 7

objectives *see also* environmental objectives; quality objectives
   CIS and 50–51
   ESDP key 102–4
   integration regime 17
   internal market regime 13
   sustainable development regime 22–3
   type of, key feature of governance regimes 7
   Water Framework Directive 27–9
obligations *see also* control; planning obligations
   reporting, River Basin Management Plan and 37
   take into account, economic instruments and Water Framework Directive 41
ODA *see* official development assistance
OECD *see* Organization for Economic Co-operation and Development
official development assistance (ODA)
   international climate change policy progress and 211
opt-outs *see also* exemptions
   EC Treaty Article 95, REACH framework and 72–3

Organization for Economic Co-operation and Development (OECD)
  'territorial capital' concept 110–11
origins
  ESDP 98–101

Pachauri, Rahendra 179
parallelism
  principle of lack of between EC internal and external competence 129
Parliament *see also* European Union
  authorization and restrictions procedures comparison and, REACH framework 77–8
  ideological orientation, factors affecting governance evolution 5, 7
  integration regime and 14–15
  substances requiring authorization and 70–71
participation *see* public participation
Plan of Action *see* Action Plan
planning obligations *see also* obligations
  Water Framework Directive and 36–7
political dynamic
  internal market regime 12
  key feature of governance regimes 7
  sustainable development regime 19–20
politics *see also* international politics
  international climate change policy progress and 185–6
  IPCC Fourth Assessment Report and 173–9
  Member States' environmental policy, environment regime and 9–10
polity
  changes, governance evolution and 5–7
pollution *see also* health
  governance evolution and 8
precautionary principle
  contestability and, REACH framework 69
prior authorization *see also* authorization
  substances 'of very high concern' and, REACH framework requirement 69–70
  substitution and, REACH framework 78
problem-solving focus
  key feature of governance regimes 7
procedure
  authorization, contestability and 75–8
  restriction, contestability and 76–8
process *see* reflection process; restrictions process
process standards *see also* standards

internal market regime 12
product market
  integration, REACH framework and 91
product regulation
  REACH framework and 58
product standards *see also* standards
  internal market regime 12–13
progress
  environmental justice and 120
proportionality principle
  external competence and, 130
  trade and environment case law and, ECJ and WTO 159–60
provisionality
  REACH framework core element 58
    authorization review 80–82
    reporting/review/revision 82–4
public information *see also* information
  factors affecting governance evolution 4
public participation
  concerns about Common Implementation Strategy and 51–4
  factors affecting governance evolution 4
  Water Framework Directive and 43–5

quality objectives *see also* objectives
  environment regime 10
  governance change limits and 25
  Water Framework and 29–30
quantitative status
  good water status and, Water Framework Directive 30, 32

rationalization
  Water Framework Directive and 29
REACH framework
  background 56–7
  contestability 58, 68–9
    authorization criteria 76
    authorization procedure 75–8
    classification and labelling, harmonized 71–2
    EC Treaty (Article 95) 72–3
    limitations 73–5
    Member State safeguards 72
    restrictions 71
    restriction criteria 77
    restriction procedure 76–8
    substance evaluation 69
    substance requiring authorization 69–71
  core elements 58

environmental protection and 91
governance approach embodied in 57–8, 90
health protection and 91
implementation phase importance 58–9, 90–91
industry responsibility 58
   authorization 66–8
   registration 60–66
information and
   generation and dissemination priority 57
   regulatory tool, as 59
product market integration and 91
product regulation and 58
provisionality 58, 80–84
substitution 58, 78–80
transparency 58
   information 84–6, 88–90
   substance classification and labelling 86–8
reasonableness
   trade and environment case law and, ECJ and WTO 160
reflection process *see also* revision
   contestability limitations and, REACH framework 74–5
regional policy *see also* integration
   cohesion and 115
   ESDP and 102–3
   spatial policy advancement and 94–6
   'Territory of the EU' and 109
registration
   industry responsibility and, REACH framework 60–66
regulation *see* control; product regulation
regulations *see* legislation
report *see also* chemical safety report
   IPCC Fourth Assessment 173–4, 176–9, 184, 203
      key messages 177–8
      preparing 174–6
reporting
   obligations, River Basin Management Plan and 37, 40
   provisionality, REACH framework 82–4
requirements
   prior authorization for substances 'of very high concern', REACH framework 69–70
   registration, REACH framework 60–63
   substance authorization, contestability and, REACH framework 69–71
restrictions *see also* limitations
   contestability, REACH framework 71
   criteria 77
   procedure 76–8
   transparency and, REACH framework 88
restrictions process
   substitution and 79–80
review
   authorization, provisionality and, REACH framework 80–82
   provisionality, REACH framework 82–4
revision *see also* reflection process
   contestability limitations and, REACH framework 75
   provisionality, REACH framework 82–4
Richardson, T. 111, 113, 115
River Basin Management Plan
   district analyses 38–40
   good water status exceptions and 34
   planning and reporting obligations and, Water Framework Directive 37–8
   public participation and 44–5
rules *see also* standards
   registration requirements and, REACH framework 62
Russia
   international climate change policy progress and 185, 202, 204

Safety Data Sheet (SDS) *see also* data; information
   access to information and, REACH framework 89
   transparency and, REACH framework 85–7
Schwarte, C. 93, 124
SDS *see* Safety Data Sheet
security *see* energy security
SIEF *see* substance information exchange forum
South Africa
   international climate change policy progress and 201
Spain *see also* European Union
   Water Framework Directive and 42
spatial strategy *see also* strategy
   ESDP and 94–8
SPS Agreement *see* Agreement on the Application of Sanitary and Phytosanitary Measures
standards *see also* rules; process standards; product standards
   disclosure of hazardous substances, transparency and, REACH framework 85

standards *(cont.)*
  registration requirements and, REACH framework 61–2
status *see* chemical status; ecological status; good water status; quantitative status
Stern Review on the Economics of Climate Change 172, 184
strategy *see also* Common Implementation Strategy; EU Sustainable Development Strategy;
  EU climate change and energy 187–8, 200
  spatial, ESDP and 94–8
Structural Funds
  reform, spatial policy advancement and 95
subsidiarity principle
  external competence and, 130
  integration regime and 14, 17
  Water Framework Directive and 38
substance
  classification and labelling, REACH framework
    contestability and, 71–2
    transparency and 86–8
  evaluation, contestability and, REACH framework 69
  hazardous, transparency and disclosure of, REACH framework 85, 88
  registration requirements, REACH framework 60–61
  requiring authorization, contestability and, REACH framework 69–71
substance information exchange forum (SIEF) *see also* information
  data sharing and 65–6
  industry accountability in registration and, REACH framework 64
substitution
  REACH framework core element 58, 78–80
supply chain
  information, transparency and, REACH framework 84–6
supremacy principle
  AETR doctrine and, 133
sustainable development
  environmental justice and 124–6
  ESDP and 106–7
  integration regime and 14, 17
sustainable development regime *see also* governance regimes
  governance change limits and 25–6
  governance evolution and 18–24

TBT Agreement *see* Agreement on Technical Barriers to Trade
TEN-T *see* trans-European transport network
territorial solidarity
  cohesion and 117
'Territory of the EU' *see also* integration
  ESDP and 92–3, 107–11, 115
testing *see also* vertebrate animal test
  industry accountability in registration and, REACH framework 63
  provisionality and, REACH framework 81
  registration information and, REACH framework 62
trade and environment
  external relations and 127–8
    common commercial policy, EC Treaty Article 133 142–4
    environmental sector, EC Treaty Article 174 142, 144, 168
    MEAs 144–6
    sector described 141–2
    GATT 146–53
    SPS Agreement 153–6
    TBT Agreement 156–7
    WTO context 146–8, 157–61
trans-European transport network (TEN-T) *see also* mobility; networks
  cohesion and 116
  environment and 114
  environmental justice and 123
  spatial planning and 96
transparency
  REACH framework core element 58
  information 84–6, 88–90
  substance classification and labelling 86–8
treaties *see also* EC Treaty; legislation
  ECSEE (Energy Community South East Europe Treaty) 169–70
  governance evolution and 5

UK *see* United Kingdom
UN *see* United Nations
UNFCCC *see* United Nations Framework Convention on Climate Change
United Kingdom (UK) *see also* European Union; Member States
  good water status exceptions and 35
  integration regime and 17
  international climate change policy progress and 190
  public participation and 45

River Basin Management Plan 40
'Territory of the EU' and 110
United Nations (UN) *see also* Bali UN Climate Change Conference
  international climate change policy agenda and 194–5
United Nations Framework Convention on Climate Change (UNFCCC) *see also* climate change; Kyoto Protocol on Climate Change; legislation
  EC and 163
  EU participation in work of IPCC and 180–81
  international climate change policy progress and 172, 183, 197–8, 203, 204, 207, 211
  international climate change regime and 161–2
  responsibility issues for Kyoto Protocol non-compliance and 165
United States of America (USA)
  environmental justice in 121–3
  ESDP comparison 93
  GATT Article XX(b) application and 148–9
  GATT Article XX (g) application and 152–3
  international climate change policy progress and
    after Bali 210
    Bali UN Climate Change Conference 201–7
    major economies meeting 192–4
  IPCC and politics and 179
  TBT Agreement and 156

vertebrate animal test *see also* testing
  registration information and, REACH framework 62

voting rules *see* decision-making mode

Wagner, Wendy 59, 79–80
water protection
  fragmentation and 29
  law and governance 27–9, 55
Water Framework Directive *see also* Common Implementation Strategy; legislation
  governance beyond 45–6
  governance techniques and
    economic instruments 40–42
    information forcing 36–40
    public participation 43–5
  main provisions 29–36
  provisionality and, REACH framework 81–2
Watson, Robert 179
welfare
  spatial policy advancement and 97–8
Witoelar, Rachmat 201
worker
  substance classification and labelling and, REACH framework 86–7
World Bank
  international climate change policy progress and 190–91
World Trade Organization (WTO)
  ECJ Opinion 128, 143–4
  international climate change policy progress and 196
  trade and environment external relations in context of 146–8
    EU behaviour in 157–61
  trade and environment sector description and 142